WORMS

by Katherine V. Nespojohn
ILLUSTRATED BY HARIS PETIE

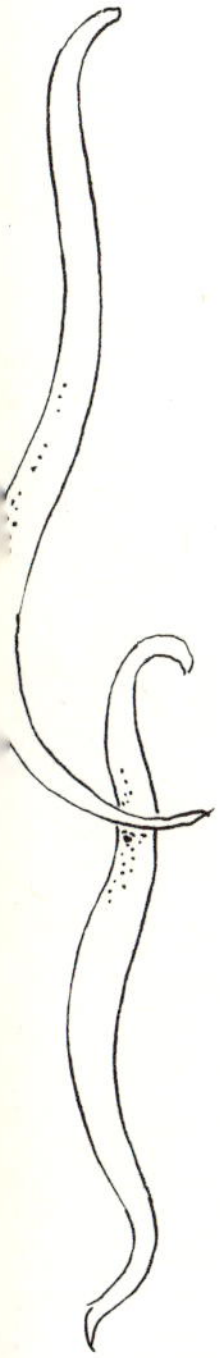

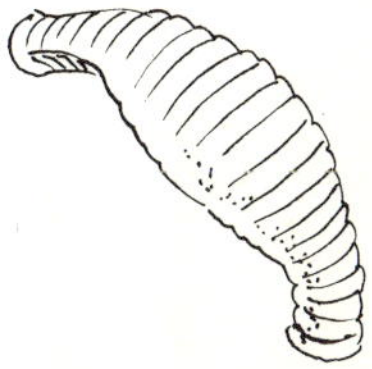

◄─ A FIRST BOOK ─►
Franklin Watts, Inc.
New York | 1972

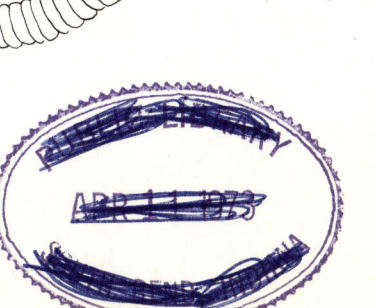

To my husband, Joseph,
and daughter, Nora,
who helped me
"worm" my way
out of household
chores while writing
Worms

Also by the author:
Animal Eyes

SBN: 531-00766-9
Copyright © 1972 by Franklin Watts, Inc.
Library of Congress Catalog Card Number: 73-182296
Printed in the United States of America
6 5 4 3 2 1

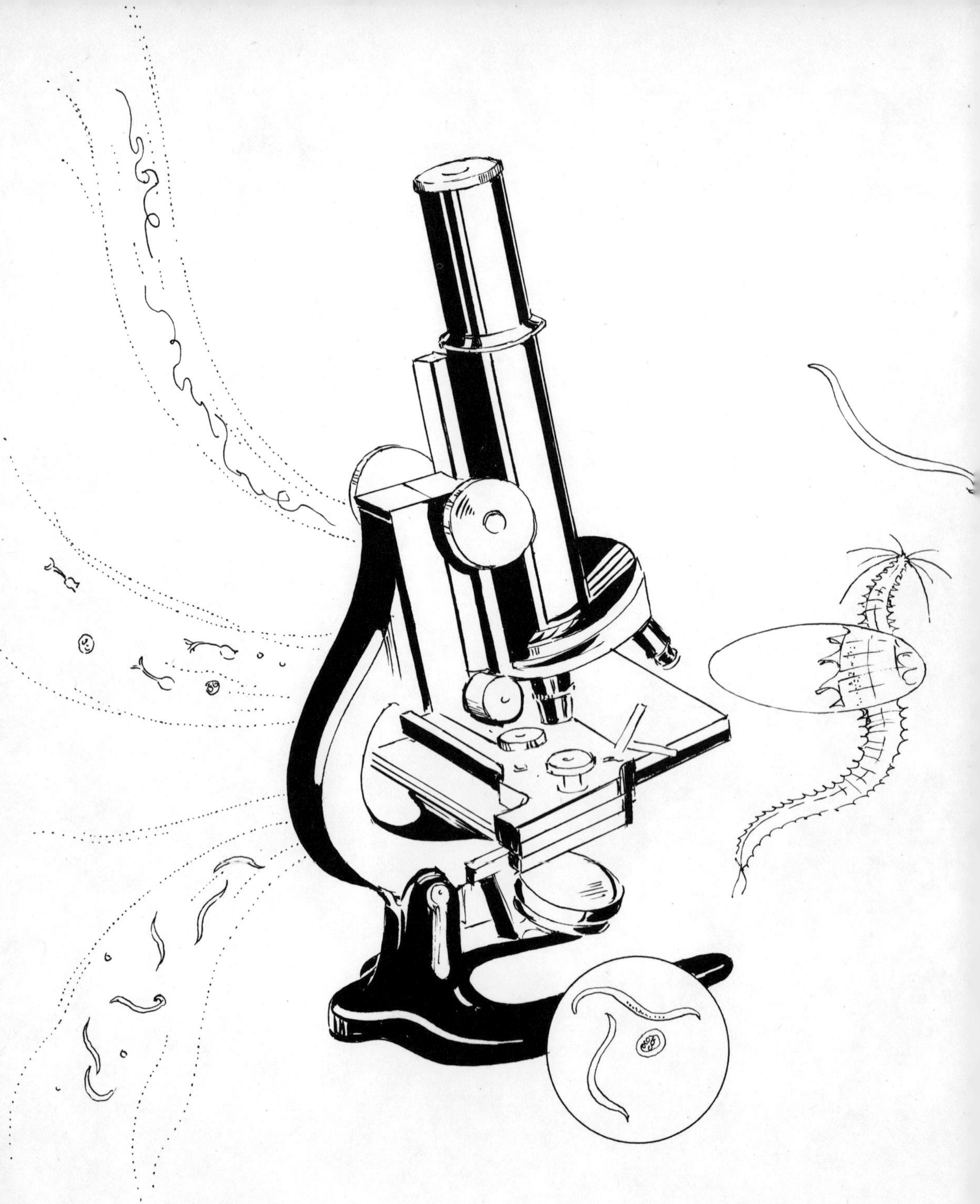

Contents

1. Worms or Not Worms   3
2. Worm-Watching   9
3. Symmetry and a Sideshow   17
4. Planaria, a Famous Flatworm   23
5. Flukes and Tapeworms   35
6. The Pesty Roundworms   45
7. Earthworms—the Underground Wonders   55
8. More Worms With Rings   67
9. Worms and Man   77
    Index   81

Worms

1. Worms or Not Worms

Worms are everywhere. They can live below you, above you, beside you, and even inside you. Each kind of worm has its own special way of life. It has evolved its own structure to meet the conditions of its homesite.

Wherever you live, by the sea, near a river or lake, in landlocked country, on a mountainside, or in a valley, it is possible to find worms to look at in your home laboratory. If the weather permits, you can start your search right away.

Anyone who wants to collect worms, and to observe and record their characteristics and behavior, might first ask the question: What is a worm?

Sometimes the word *worm* is given incorrectly to any small creature with a long, soft, slender body that crawls or slithers on its underside.

Is that really a worm in the apple? When you sing "shine, little glowworm" are you singing to a worm? The inchworm, or measuring worm, seems to walk in loops — but is it a worm? Do you blame worms for munching holes in last year's favorite sweater? What are mealworms, silkworms, or ringworm?

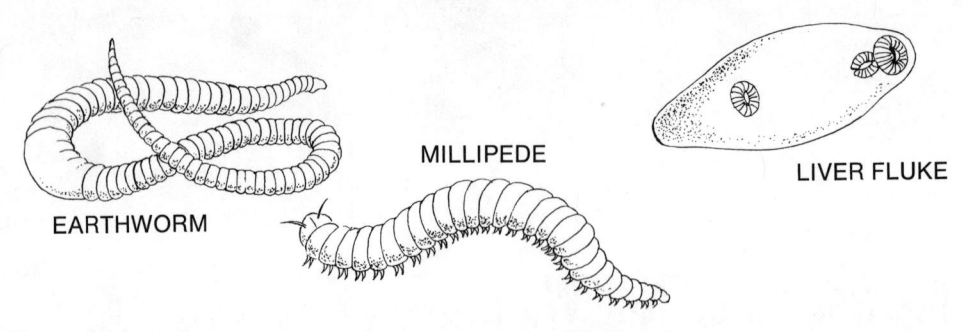

Even though it may look like a worm, the millipede, as well as the centipede, belongs in the phylum Arthropoda, the jointed-legged animals.

More than two thousand years ago, Aristotle, a Greek philosopher and scientist, saw the need to put living things in order. He described specimens and then placed them on his "ladder of nature." Man sat on the top rung. This incomplete, and sometimes inaccurate, listing was used until about 225 years ago. At that time, a Swedish physician and naturalist, Carl von Linné — better known by his Latin name, Linnaeus — created his binomial system.

"Binomial" means that each specimen was given two names. Since it was important to use the language of scholars everywhere, it was natural to choose Latin. Thus, man was called *Homo sapiens* — *homo* meaning "only one of its kind," and *sapiens* meaning "wise." (Greek belongs to the same family of languages as Latin, so Greek terms are also used in classification systems.)

All living things, including the forty thousand worm species, are placed in groups known as phyla. One group is called a phylum.

Flatworms are in the phylum Platyhelminthes. (The Greek word *platy* means "flat," and *helminthes* means "worms.") The

COMMUNITY BAPTIST CHURCH
LIBRARY

phylum has three subdivisions, or classes: Turbellaria, or free-living (not attached to anything else), flatworms; Trematoda, or flukes, which are parasites, living inside or outside other animals; and Cestoidea, tapeworms that live as internal parasites when in the adult stage.

Roundworms are in the phylum Aschelminthes, which includes many classes. The most important class to man is Nematoda (*nema* means "thread"). About fifty different species of this worm have a favorite hideout — man!

Segmented worms are found in the phylum Annelida. They are identified by bodies composed of rings or segments. (The Latin word *anellus* means "little ring.") The phylum subdivisions, or classes, include: Oligochaeta, the earthworms and their relatives; Polychaeta, the sandworms and tube worms; Hirudinea, the leeches; and Archannelida, other marine worms.

Three classes of the phylum Platyhelminthes: l. to r.: Turbellaria, Trematoda, and Cestoidea.

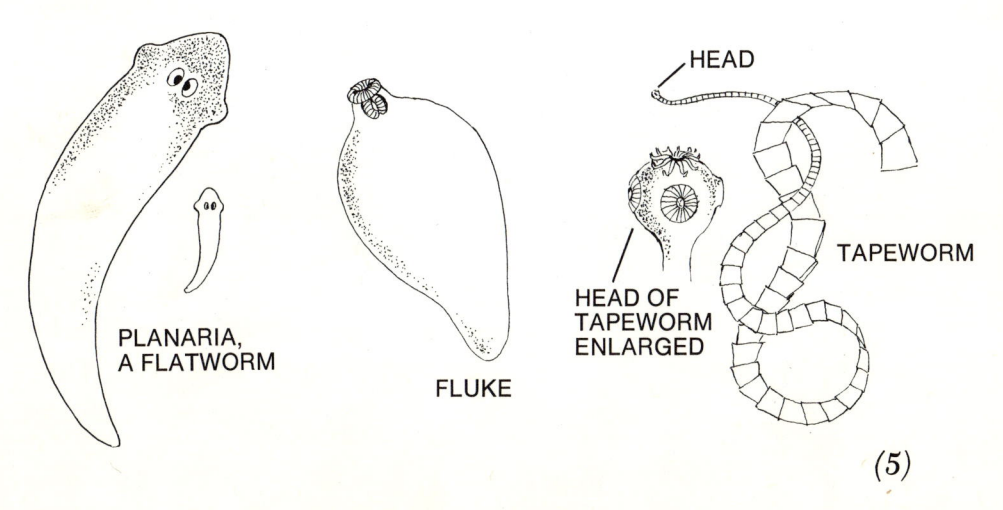

As you get to know different members of each phylum, you will begin to appreciate the process of classification. The structure and the function of the organisms are the keys that unlock the door to each phylum.

You have already done some classifying if you have arranged your clothes closet. Certain things are assigned to shelves, hooks, hangers, and boxes. There are various classes of clothes: woolens or nylons, warm or cool, dark or light, old or new, sporty or dressy. Housekeepers and storekeepers must be masters in the art of classification or they would never be able to find anything.

It might seem safe to say that man has seen almost every living animal on his planet, and that all existing creatures can fit into one of twenty-seven phyla. But this is not true. Recently, for instance, scientists said that a newly discovered tiny marine worm would not fit into any existing phylum. A new group was suggested: the phylum Gnathostomulida. The name is almost six times longer than the worm!

Again, let us ask: What is a worm? Part of the answer is that a worm is an animal without a backbone, an animal that crawls or slithers. Worms are generally, but not always, long and slender. However, a more detailed answer could be found by answering the separate questions: What is a flatworm? What is a roundworm? What is a segmented worm?

Are these creatures worms?

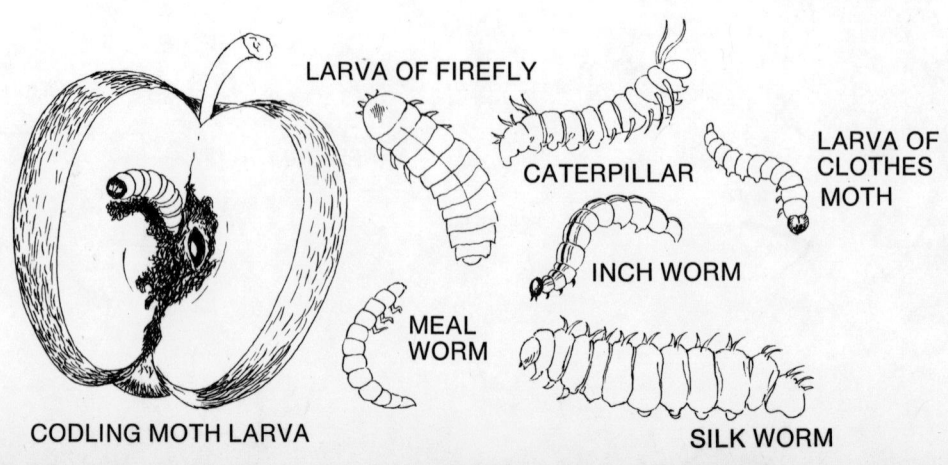

LARVA OF FIREFLY

CATERPILLAR

LARVA OF CLOTHES MOTH

INCH WORM

MEAL WORM

CODLING MOTH LARVA

SILK WORM

All worms go through some form of reproductive cycle, which ends with the formation of the adult worm. Many other invertebrates — animals without backbones — have phases in their life cycles in which they resemble the worm. Eventually, however, other changes take place before the adult emerges.

This is true of the "worm" in the apple. It is actually a larva (early stage) of the codling moth. The glowworm is the wingless female larva of the firefly-to-be. Most caterpillars are living a temporary existence. The animal's adult stage will be in the form of a moth or a butterfly. The inchworm becomes a moth. The "clothes moth" ruined your sweater, but it was in the larval stage at the time — a wool-eating, wormlike phase. The mealworm changes into a beetle, and the silkworm into a moth. Ringworm is not a worm, but the name of a skin disease caused by a fungus, a simple plant.

Worms live largely hidden lives. It has taken patient probing into a dark, wet, muddy, but interesting underworld to expand man's knowledge of these important creatures.

2. Worm-Watching

You are about to become a discoverer, an explorer in the world of worms. By observing, collecting, questioning, recording, and thinking, you may experience the real thrill of new knowledge — discovery.

The tools needed for worm-watching are fairly simple. They include a small shovel or digging fork, assorted containers and plastic bags, some scraps of raw liver or beef, string, a magnifying glass, a flashlight, a thermometer, a ruler, a notebook, and pencils. However, the most important tool is curiosity.

Worms have no protective outer covering. They have no exoskeleton containing burrs, spines, scales, or bonelike plates. Therefore, for the sake of survival, they have been forced to develop an underworld of their own. It is in this obscure, unexplored, under-the-surface environment that worm-watchers must do their sleuthing.

Armed with your equipment, start probing a shady spot of rich, moist topsoil. With the fork, gently loosen a section of sod to about 8 inches in depth. Carefully flip the grassy section over, and be quick with your eyes. If an earthworm has been disturbed, it was alerted to your presence by the vibrations set up in the soil. The animal will quickly try to disappear.

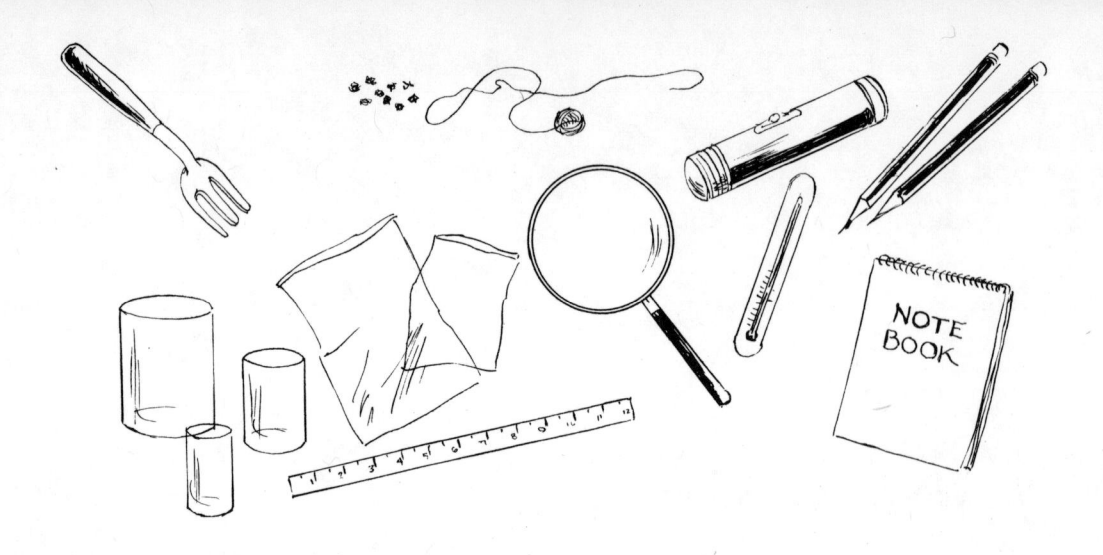

Now probe the hole deeper and deeper, and take mental notes when the next worm is uncovered. Does it appear to be disturbed by the intrusion? Does it crawl on one body surface, or does it roll about using any part of the body wall? Does the worm move sections of its body or its entire length at one time? What other questions occur to you?

Since outdoor laboratory work is sometimes more difficult than that done indoors, you might set up a checklist as an observation guide. Sometimes, conditions do not allow for note-taking, but check marks are easy to record. The sample "Worm Observation Sheet" on pages 12 and 13 will help you to get started. In time, you will probably create your own data sheets.

If you set up your outdoor laboratory in an open field or a section of your own yard, many specimen notes may be recorded on a single sheet. If the work is done on different days, the headings of the chart must show that. In time, if the data are carefully recorded, some findings may be drawn from a single page. For example: if seven earthworms were observed, and seven check marks appear on the line for "dug in deeper," you may be on the verge of a discovery. (You may find that most earthworms try to disappear into new underground tunnels.)

The time has come to start your worm search.

Before leaving the outdoor observation site, capture a few worms and place them in a shallow flat container with a little moist soil. Using the magnifying glass, study the head. Do you see eyes or ears? Can you find a mouth on the forward end? Take notes on the way the worm crawls — using, it seems, all its muscles at once, contracting and expanding its length. About how long is each specimen when it is fully stretched out?

Hold the worm, and let it slip through your fingers. Do you feel bristlelike structures on the lower surface? Use the magnifying glass to study these, and try to guess how the worm makes use of them. What other openings or marks can you find on the body wall?

Worm-watching can be done in water as well as on land. However, at a water site it is easier to first collect the animals and then study them indoors.

A common freshwater flatworm called planaria lives in cool running streams, clear ponds, or lakes. Tie a cube of raw liver or beef on a string and hang the "trap" low in the water, so that it rests near a clump of plants, some rocks, or an old log.

After about half an hour, pull in the string and swish the bait through some pond water in a pan. (A white enamel pan is the best container for these worms.) Wash the tops of some of the plants in the pan. The worms, if any, will look like tiny brown

Worm Observation Sheet

Date _____
Time _____

Observation site _____
Outdoor temperature _____

A. *Location:*

Specimen No.

	1	2	3	4	5	6	7	8	9	10
Moist soil										
Dry soil										
Rich soil										
Sandy soil										
Depth in soil										

Notes, comments, questions:

B. *Movements:*

Toward light							
Toward dark							
Stayed exposed							
Dug in deeper							

Notes, comments, questions:

C. *Body Notes:*

Length							
Color							
Any appendages							
Marks on body							
Skin, moist							
Skin, dry							
Skin, rough							
Skin, smooth							

Notes, comments, questions:

balls. But if you watch, the balls will stretch out and you will see worms that have pointed tails and arrow-shaped heads.

Keep the magnifying glass handy for instant observation. Work patiently until you have captured about twenty specimens for indoor study. If you wrap some liver in wire screening or place it in a covered wire strainer, you can leave it in the stream overnight. The wire will keep unwanted animal burglars from stealing the bait.

House plants can be a source of roundworms. Look in the water that has seeped through the soil and has been standing in a container under the plant. In bright light, examine the seepage with a strong magnifying glass or a small microscope. The worms will look like transparent threads — the most active threads you will ever see. Keep a collection of a house plant's surplus water for later study.

Roundworms are found at low tide in salt water or on freshwater lakeshores. Choose a spot that harbors large amounts of dead or decaying plant material. Place a scoopful of bottom muck and some clear water in a container. When the water clears, look for threadlike mini-whips that seem to be endowed with perpetual motion.

The most familiar worm in a water environment is the fisherman's free bait, the sandworm. Using the fork, turn over heaps of

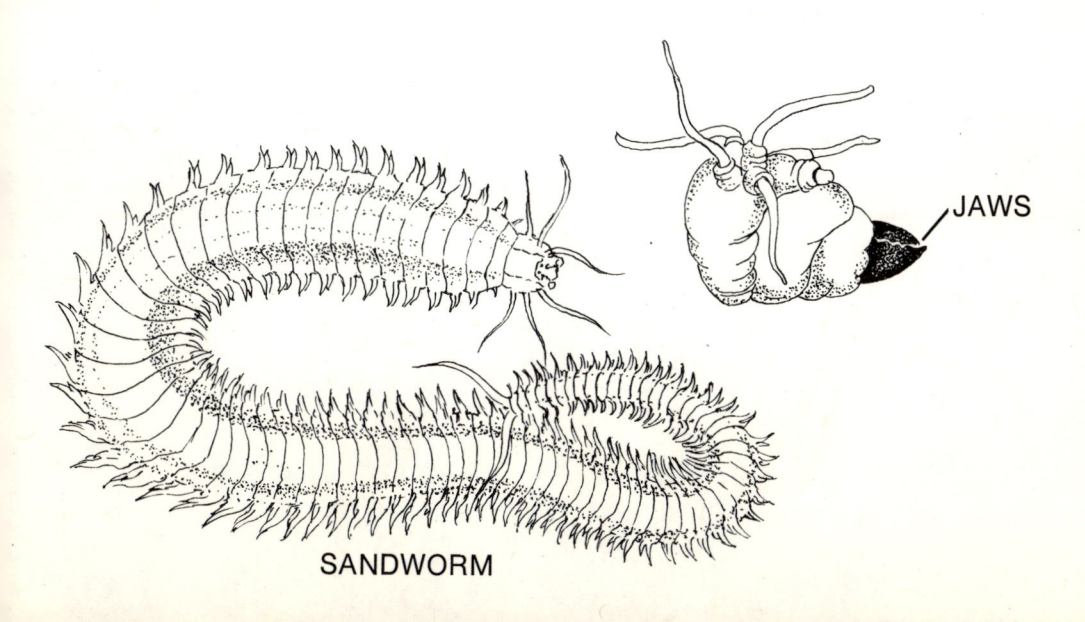

JAWS

SANDWORM

wet sand containing quantities of decaying organic muck. Unless the worm's entire underground hideout is excavated, the animal will quickly burrow out of sight — so dig down deep. Use the fork or shovel to place sandworms in a container. The worm's head is equipped with horny-toothed jaws that may nip an unprotected hand. Cover your collection with wet sand and seaweed for indoor study.

Successful worm-watching depends on favorable weather conditions, for the weather affects the soils, the water, and the life cycles of the animals. Whether your efforts are rewarding each time or not, you are learning about creatures that most people ignore.

3. Symmetry and a Sideshow

All animals have some basic body shape. If the shape can be divided by an imaginary line into equal and similar parts, the body has *symmetry*. Irregular forms, such as in some one-celled creatures, do not have symmetry. They are *asymmetrical*. They cannot be divided into equal and similar parts.

Study the symmetry of any objects that are near you — a vase, a chair, a tree, a bird, and even you! Which is easier to find, regular or irregular shapes? Does water or milk have shape? You are correct if you said that liquids take on the shape, or form, of their containers.

There are three types of symmetry: *spherical, radial,* and *bilateral*. Most animal shapes fit one of these patterns. Some Protozoa, or one-celled creatures, are nothing but microscopic balls of protoplasm. They exhibit spherical symmetry, at least while at rest in a globelike shape. A line cut through the center in any direction divides such an animal into two equal half *spheres*, with no right or left sides, no top or bottom. They may become asymmetrical as the protoplasm in their single cell spreads out in several directions at once.

ASYMMETRICAL　　SPHERICAL　　　　　　　　　　BILATERAL

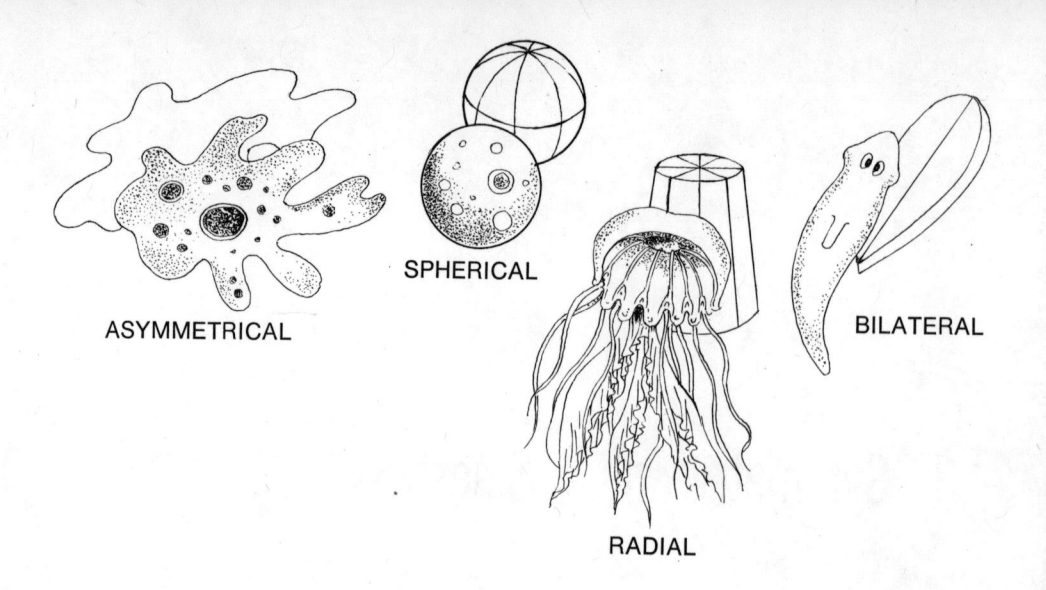

RADIAL

If the spherical animal has a mouth surrounded by hairlike-fringe or mini-tentacles, a lengthwise cut through the center of the mouth in any direction will divide it into similar halves. This illustrates radial symmetry. Think of an orange with a thumb-size hole, or "mouth," at one end. You can see that this symmetry is different from spherical symmetry. The starfish and some vaselike sponges are examples of this type of body plan. Do they have a right and a left side?

Animals with a definite head end (anterior) and rear end (posterior) are referred to in terms of their upper (dorsal) and lower (ventral) surfaces, as well as by their right and left sides. These terms apply to animals that have bilateral symmetry.

"Bilateral" means two sides. But it tells you that there is only one line from anterior to posterior that will cut the animal into similar halves. That cut must be in only one direction, from the dorsal (upper) through to the ventral (lower) surface.

You have bilateral symmetry. Stand in front of a mirror and

you will see that it is true. You will also see that single organs, such as your nose and mouth, are located in a midline and lie in the exact path of the proposed cut.

What kind of symmetry does a bird have? A jellyfish? A dog? A goldfish? (You are right if you said bilateral, radial, bilateral, and bilateral.)

What does symmetry have to do with worms? If you work in any area of science, there are certain terms that must be understood. Symmetry is one way of finding out how complex an organism is. It is always referred to in listing the general characteristics of a species. It has been a clue in classifying all animals, from the simple to the complex.

Exploring the worm world is somewhat like going to a circus sideshow outside the big tent. As you wander about, you get a glimpse of great things to come. The sideshow is set up to make you so curious that you want to see more.

"Out of sight, out of mind" seems to be the worm's key to survival. So, most people have seen very few of these creatures. The common ones are reddish or muddy brown. But worms can really be quite colorful. Wander around and see for yourself.

Free-living flatworms called polyclads look like 2-inch-long leaves. Some are varicolored or attractively striped, with alternating dark and light colors. Polyclads and their relatives, called acoels, are marine worms. Acoels are seldom seen since they hide their ⅛-inch-long bodies under stones and in algae on the shallow sea bottom. They are usually white or a very pale color. But one species called *Convoluta* is brilliant green — so green that masses of them tossed above the high-water mark seem to splash green stains everywhere. The worm is colorless when hatched, but it becomes green as it takes into its body one-celled animals known

DETAIL OF HEAD

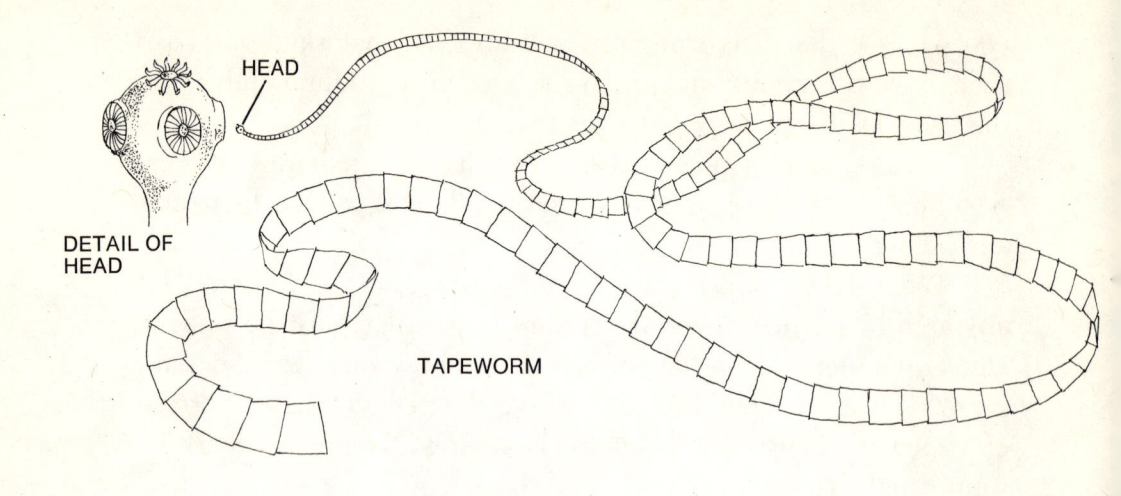

HEAD

TAPEWORM

as flagellates. These animals contain chlorophyl, and they need sunlight to survive.

The world of worms has its own version of the fat ladies, tall men, and midgets in the sideshow. Sizes in worm families stretch from the microscopic to hundreds of feet long. Sometimes it is difficult to estimate a worm's length. Some worms can contract their bodies like an accordion. You will probably discover this.

Tapeworms, a form of parasitic (living on others) flatworm, probably hold the world's worm-length record. Beef tapeworms have been measured at 100 feet from end to end. One recorded worm tale claims that a lady once housed in her intestines six fish tapeworms. The combined length was almost 300 feet.

Australia boasts the world's largest earthworms. They are 4 feet long and as thick as three or four times the width of your thumb. One worm was reported at 11 feet long. These worms have a high protein content. Some Australians fry them in hot fat for a very special meal.

Every sideshow has a fire-eater. The closest thing in the worm world is a specimen with a quality known as *luminescence*. Some marine worms actually produce light, especially when their bodies strike objects. Ripples of violet light have been seen to flash out, but no one seems to know how the light helps the worm. Perhaps the luminescence is simply ornamental, not useful at all.

Each characteristic makes an animal fit into one of many structural patterns. This is what makes the sideshow possible.

4. Planaria, a Famous Flatworm

If you were asked to name an animal used in scientific experiments, you would probably choose the rat or the guinea pig. However, one of today's most fascinating research specimens is a flatworm called planaria.

If you cannot find live planaria specimens, you can buy some from a biological supply company. Perhaps a science teacher may be willing to help you get some. Your interest in these "aquaworms" will more than triple if you can work with them as they twist and turn through their watery environment.

Anatomy is a good starting point to study any animal. It is the body structure that determines the organism's function, its habitat, and its relationship to man.

Planaria is just one of about ten thousand species in the phylum Platyhelminthes. But its characteristics are common to all members of the class Turbellaria.

This group has bilateral symmetry, a body style with a definite head, or anterior region, which performs the leading role in locomotion, and a posterior region, which tapers off to a point. Planaria's ¼- to ½-inch gray or brown body is flattened dorsoventrally, or from top (dorsal) to bottom (ventral). The ventral surface, which is covered with cilia, or hairlike fringe, tends to maintain contact with the surface on which the worm slithers.

Although it is free-living, the planaria does not swim. It glides on surfaces. If no suitable surface is available, the animal secretes its own magic carpet — strings of mucus, on which the cilia can get a hold and move the body forward.

Anyone seeing planaria for the first time will probably say, "It's cross-eyed!" And that's exactly what it is. The head has at least one pair of primitive eyes, *ocelli*, which are black pigment cups filled with photosensitive cells that connect directly with the brain. Flatworms do have brains — very simple nerve centers. The black area of each eye cup is located toward the center point of the head, creating the cross-eyed effect.

The eye pigment shades the cup from light, except light coming from one direction. The body's "skin" layer directly over the eye has no pigment (color), so light can pass through to the sensitive area beneath. Although planaria is very responsive to changes in light, the ocelli cannot take all the credit. When the eyes have

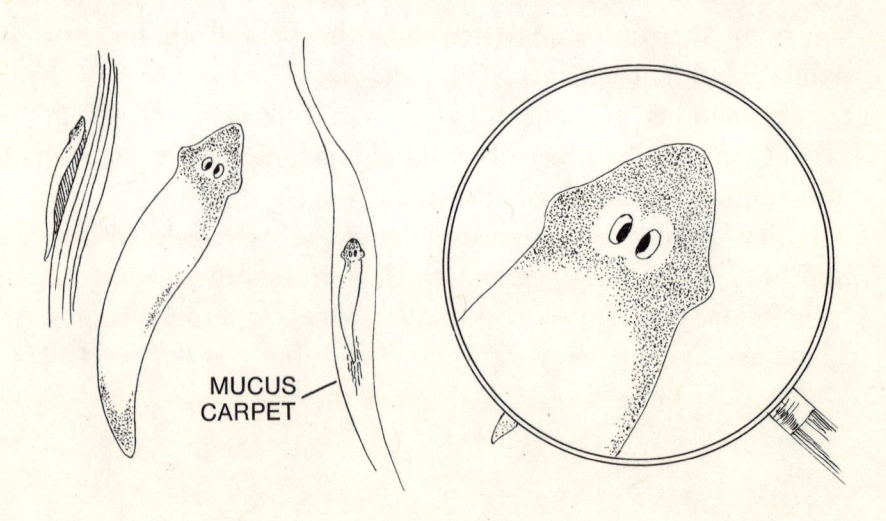

MUCUS CARPET

been removed from some specimens for scientific study, the animal still shows an awareness of light, although it responds a bit more slowly than normal. A beheaded animal also moves more slowly, but its reaction to light is not changed. Thus, it is believed that the entire body may be light sensitive.

Since the eye has no image-forming apparatus — no lens, no retina — it is not possible for planaria to see objects. The worm lives in a world of shadows. Cave-dwelling planaria have no eyes at all. But their other keener senses make up for it. Some members of the class Turbellaria, such as the land planaria, seem overequipped. They have multiple pairs of eyes, always located in the head region and near the brain. These eyes are more complex, and maybe dim images can be seen. It seems that eye structure is related to the animal's needs.

Observation is the best tool to use in learning about an animal's anatomy and behavior. Put a few specimens of planaria in shallow, clear water, and, using a magnifying glass, sketch the head region of one worm. Put in every detail.

Since eyes and ears seem to belong together, where are the ears? The lobes on each side of the head are called *auricles*. They carry brain-connected sensory cells, but they are tuned in for touch, water currents, food, and chemicals. Sound, as we know it, is not involved.

Touch one auricle with a toothpick and then the other, and record the animal's reaction. Blow through a straw so that air strikes the edge of an auricle. Does the animal move its head toward or away from touch, toward or away from air current? Which reactions would you call positive (toward) and which negative (away from)?

Now take a medicine dropper and direct a water current

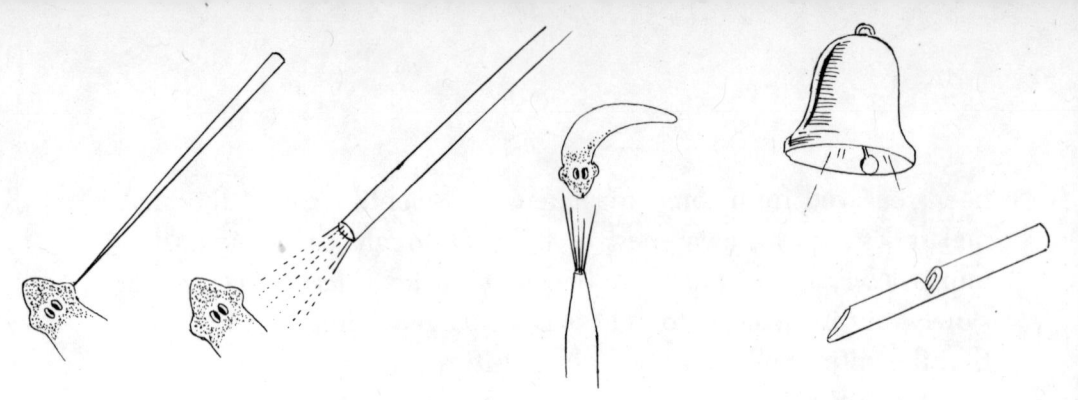

Planaria, a flatworm, is used for scientific research.

toward the head. Is the worm positively (toward) or negatively (away from) conditioned to water currents? Many trials are necessary before you can say with assurance that something is true of an animal's behavior. If your friends work with you, your data will be more impressive in a shorter period of time.

You do not have to believe any statements just because you see them in print. Test them. For example, using a bell or whistle, try to see for yourself if planaria will react to sound. Hold the noisemaker some distance from the animal, so that the head is not influenced by a shadow you may cast or an air current from your mouth. Get rid of every factor except the one you want to test.

The mouth of the planaria is not in the head, where you might naturally expect to find it. It is along the mid-ventral line, about halfway down the body's length. It is a simple circular opening, capable of changing size by muscular action. It is provided with a *pharynx*, a tube that reaches out through the mouth opening to grab food particles and to direct them into the intestine. Watch with the magnifying glass and you will see the pharynx in action — especially if you put a pinhead-size bit of raw meat in the water. The next time you see an elephant grabbing at food with its trunk, remember that a worm can do it, too.

MOUTH
PHARYNX

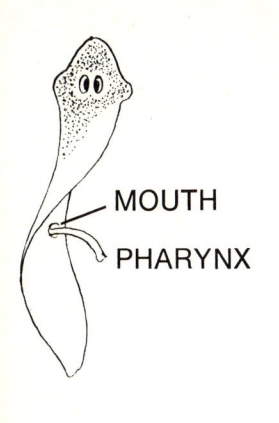

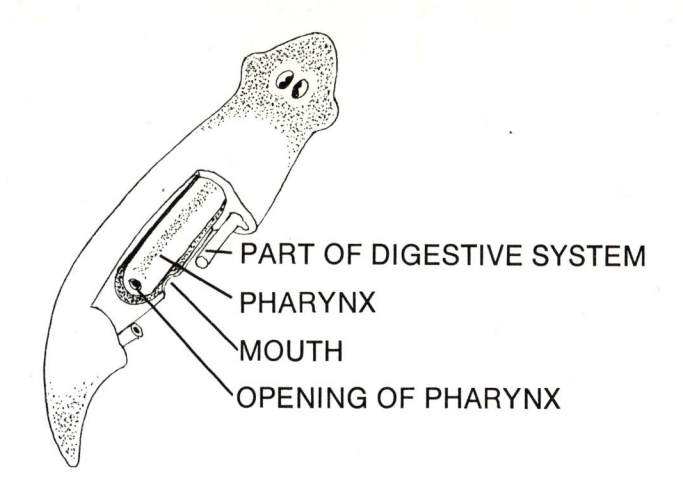

PART OF DIGESTIVE SYSTEM
PHARYNX
MOUTH
OPENING OF PHARYNX

Does the worm seem immediately aware of the food? One observer claims that when food is present, the worm pauses, swings its raised head about, turns its body toward the food, and slithers toward it in a direct line. Test the accuracy of these observations with your specimens. Watch for the animal to secrete some slime around the food, making it easier to slide it into the pharynx. Although the mouth is not located in the head, if the auricles are removed, the food-detecting ability of the worm doesn't work as well. So, it is suspected that the head tests the food, decides if it is suitable, then activates the feeding mechanism of the pharynx. Who said a worm is a simple animal?

Three suggested observation check sheets are shown on pages 28, 29, and 30. You will probably develop your own data-collecting system as you get more experienced.

The organ systems of the class Turbellaria are similar. The life processes of one worm provide information on many other related worms.

The most exciting feature of the planaria is its power of re-

Planaria and Light: Behavior Check Sheet

Test A. Worm is given choice of light or dark environment in a shallow glass dish. Cover about half of the top of a dish with black paper. For each trial, the paper is moved to expose the worm to light.

Trial Number
(Check indicates one response)

	1	2	3	4	5	6	7	8	9
A.—Worm moves to or remains in LIGHT area.									
B.—Worm moves to or remains in DARK area.									

Notes, comments, and questions:

Planaria and Light: Behavior Check Sheet

Test B. Flashlight beam is reduced to a pinhole of light by using a black opaque shield. Thin beam is directed at the head of worm. Work in a dark room gives more striking results.

Trial Number

	1	2	3	4	5	6	7	8
A. Positive head reaction.								
B. Negative head reaction.								

Notes, comments, and questions:

Planaria and Light: Behavior Check Sheet

Test C. (variation of Test B): Use same light as in Test B.

	To Right	To Left	Up	Down	None
Trial Number	1 2 3 4	1 2 3 4	1 2 3 4	1 2 3 4	1 2 3 4
1. Beam on RIGHT side.					
2. Beam on LEFT side.					
3. Beam on DORSAL center (head).					
4. Beam on VENTRAL center (head).					

5. Beam on POSTERIOR.

	Anterior Reaction			Posterior Reaction		
	Neg.	Pos.	None	Neg.	Pos.	None
Trial Number	1 2 3	1 2 3	1 2 3	1 2 3	1 2 3	1 2 3

Notes, comments, and questions:

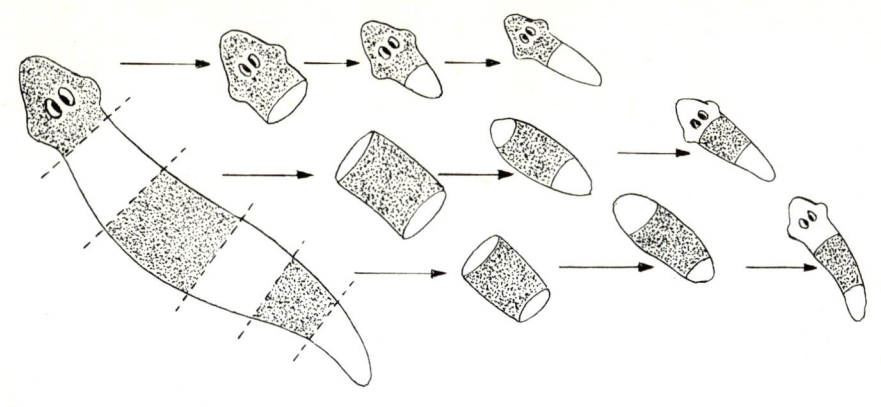

Regeneration of planaria

generation. Pieces of the animal regrow in about four weeks to form complete worms. But the head controls the pattern of regrowth. Heads that are cut lengthwise will regenerate the missing half. Several such cuts will result in as many new heads — all with the same posterior section. The old saying "Two heads are better than one" could be a planaria's proverb.

Planaria's head can regenerate many new ones after being cut lengthwise several times.

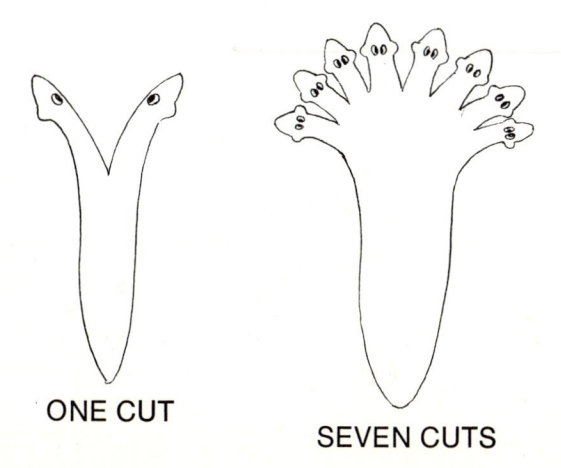

ONE CUT   SEVEN CUTS

(31)

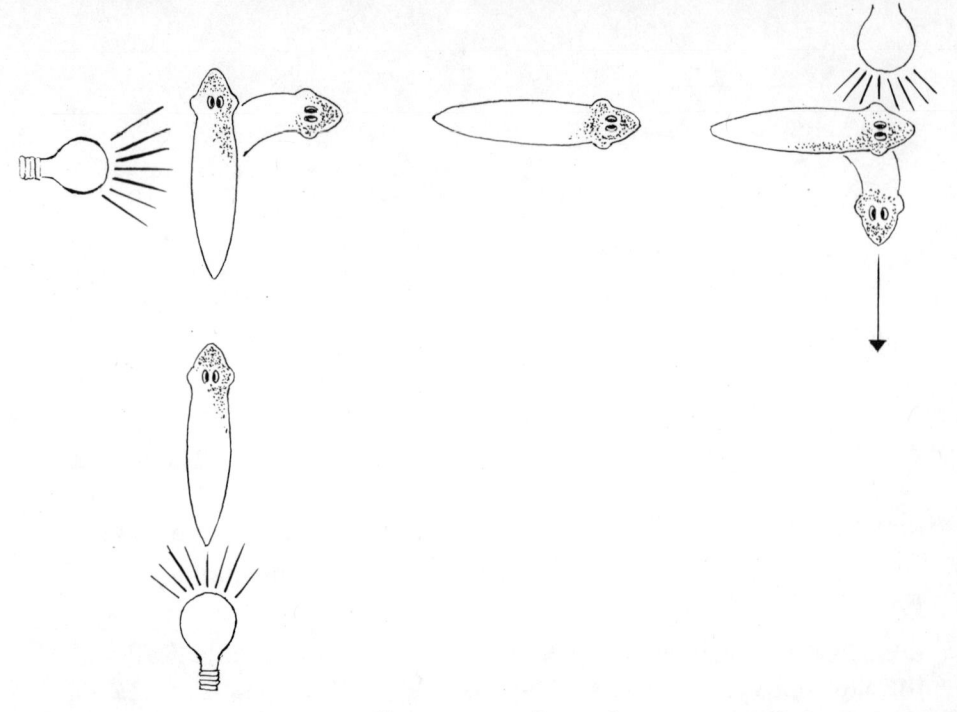

You can test planaria's responses to light. Try this.

This worm has one other exciting characteristic — it can learn. At least, man has successfully conditioned it by using the reward or punishment technique. Scientists Robert Thompson and James V. McConnell, working at the University of Texas, set up a kind of "school for planaria." Worms were conditioned with bright lights and electric shocks. They responded to light by stretching their bodies. The shock, which was given immediately following the light treatment, produced a body contraction. After about 100 trials, the worms, on being flooded with light, skipped the stretching reaction and contracted their bodies. They were, it is suspected, associating one thing with another. A worm was considered trained if it responded by contracting only, in twenty-three out of twenty-five tests.

Later, trained planaria were chopped up and fed to untrained specimens. The "cannibals" seemed to have learned by eating. They had inherited, by ingestion, digestion, or by some unknown process, some of the training that they themselves had never had. Dr. McConnell has written about memory transfer through cannibalism, but there is still much to be done with this unique idea. It doesn't seem practical to have to eat your professor in order to get a top grade on an exam!

Many questions are unanswered. Is RNA (ribonucleic acid), which is a molecule related to protein production within the living cell, responsible for memory and the transfer of learning? Can man, through RNA, help the aged who are afflicted with loss of memory? Can a long-suffering student absorb knowledge and retain it through a capsule of some yet-to-be-discovered memory material?

Maybe the schools for planaria will change the schools for man!

5. Flukes and Tapeworms

So naturalists observe a flea
Hath smaller fleas that on him prey;
And these have smaller still to bite 'em;
And so proceed, ad infinitum.
                    (Author unknown)

It is impossible for any creature to be completely alone. Animal parasites are uninvited guests sharing the good and bad fortunes of their fellow organisms. Even they — the parasites — have guests.

Two classes of Platyhelminthes depend on other animals for their lives: Trematoda, flattened, oval, leaf-shaped worms known as flukes, and Cestoidea, ribbon worms or tapeworms.

The Greek philosopher Aristotle was fascinated by worms that live in man. He wrote, "In all natural objects, there is some marvel, and if anyone despises the contemplation of lower animals, he must despise himself."

Trematoda (from the Greek *trema*, meaning "hole") is a class of about three thousand species of flatworms, commonly called flukes. They are small creatures ranging in length up to 2 inches,

(35)

depending upon the species. All flukes have oval, flattened, bilaterally symmetrical bodies with no definite heads. However, each fluke has a head end (anterior) marked by a tapering, rather blunt, point. Its method of attachment is located there.

The anterior end makes one thing very clear: it is ready to hang on after it gets into a desirable environment — usually the body of a vertebrate, such as man. It will use either muscular discs, suckers, claws or hooks, or a combination of these.

Some simple flukes live as exterior parasites. One species attaches itself to the skin or gills of fish and feeds on the host's tissues. If they find that the free ride is too swift or difficult, they crawl into cavities, such as the host's mouth. Many other aquatic animals, including amphibians and reptiles, are fluke-infested.

You may never see a fluke because they seldom make a public appearance. However, their negative importance cannot be ignored. It has been said that during World War II this worm actually helped to determine American military strength in the South Pacific. Sick soldiers suffering from these worms, an illness that was difficult to control in the field, delayed many proposed plans.

Flatworms hang onto their hosts with a variety of hooks and suckers.

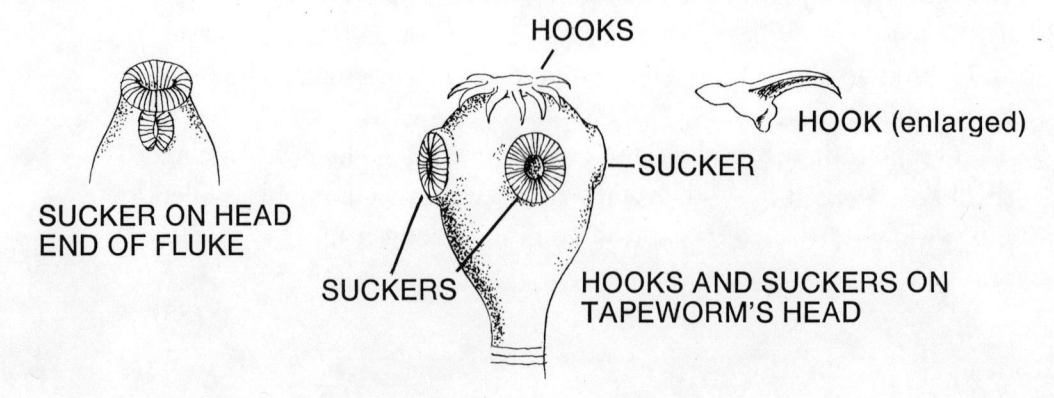

SUCKER ON HEAD END OF FLUKE

HOOKS

SUCKER

HOOK (enlarged)

SUCKERS

HOOKS AND SUCKERS ON TAPEWORM'S HEAD

FLUKES ON FINS OF FISH

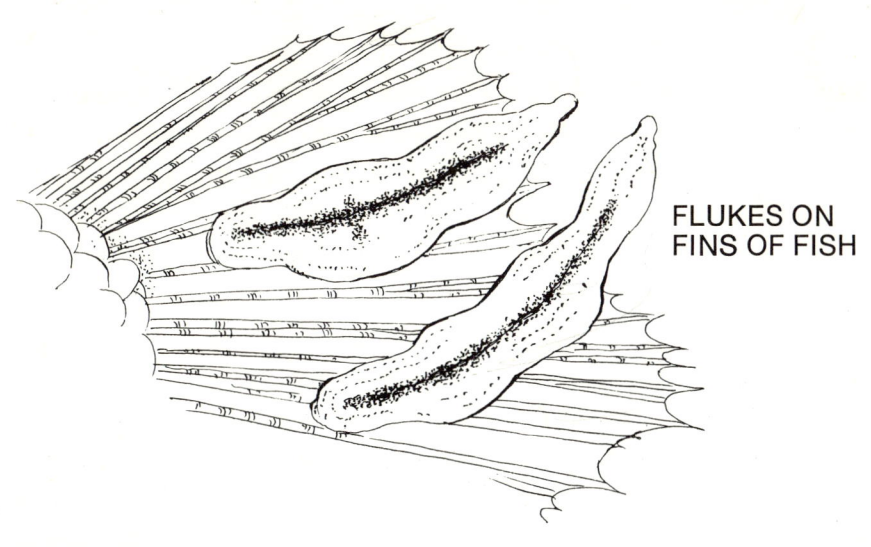

Parasitic flukes get free food and transportation.

One species, the blood fluke, infests masses of people in the Orient, Arabia, Cuba, the West Indies, and much of Africa and South America. Historians have blamed this fluke for slowing down cultural and economic progress in about one-tenth of the world's population. Schistosomiasis, or fluke disease, affects about two hundred million people. It is pushing malaria out of first place as man's greatest tropical scourge.

In some countries, the construction of canals, dams, and irrigation ditches helped to spread blood fluke infestation. But the idea that this worm is a pest only in poverty areas or in those areas with poor sanitary conditions is not true. "Swimmers' Itch" is a complaint of visitors at even the most exclusive lakefront resorts. The larval fluke is responsible for this annoyance.

Although prevention is the best cure, getting rid of worm guests is possible. If the worms have already invaded the victim, a doctor must be consulted. Prevention, diagnosis, and treatment are all related to the knowledge of the life stages through which these animals progress, from ova to adult. So, it is important to understand these life stages.

Traces of the blood fluke have been found in three-thousand-year-old Egyptian mummies, but the life cycle of the pest has been known for only about sixty years. The term "host" means the animal that houses the parasite. The term "intermediate host" means that at least two hosts are necessary to complete the cycle.

Many flukes need a species of tropical snail as an intermediate host. Since man acts as the final host, human cases of this illness occur only in regions where the "middleman" snail exists. American-bred infections are rare, even though some of the tropical snails live here. Most fluke infestations come to this country as stowaways inside tired tourists.

Human blood flukes use man and snails as hosts.

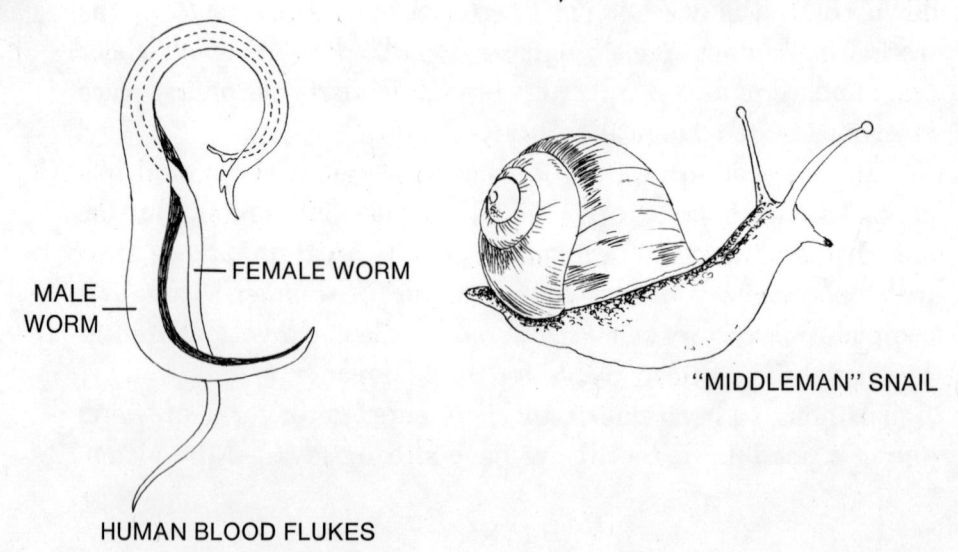

MALE WORM — FEMALE WORM

"MIDDLEMAN" SNAIL

HUMAN BLOOD FLUKES

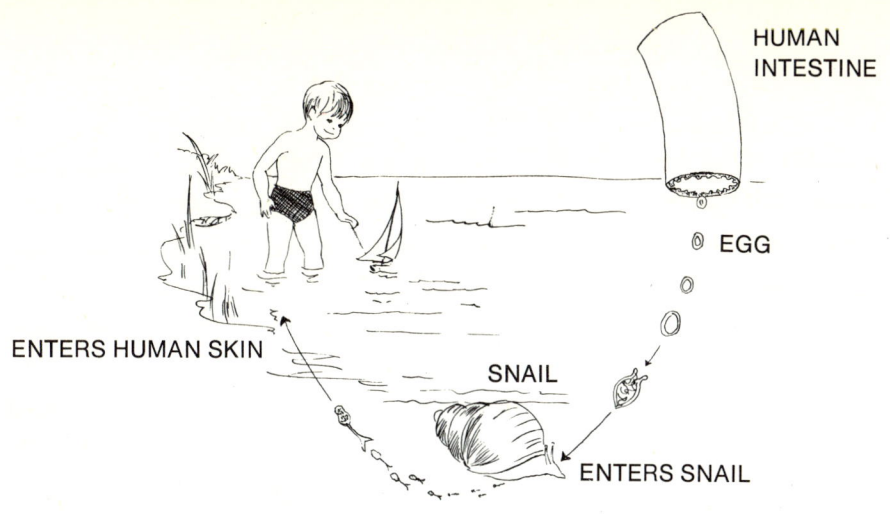

Life cycle of human blood fluke

    The worms that annoy man are identified by their favorite human living quarters. There are three species of blood flukes: liver, intestinal, and lung.

    Fluke life cycles are similar. If you know about the blood fluke, you pretty much know the life story of all flukes. In the human (or final) host, the adult male and female (other flukes are *hermaphroditic;* that is, each fluke contains both sexes) live together in the veins of the bladder or colon. They produce fertile eggs. The sides of the male's body fold over to form a groove, into which the female's long, slender body can fit during the mating process. The worms cling to the host with suckers. They feed on his blood, causing him to complain of body pains, fever, and violent dysentery. Some human hosts have been known to house the parasite for as long as thirty years, resulting in extreme weakness and eventual death.

    The embryos in the parasite's eggs secrete a toxic material. It penetrates the eggshells and the capillary walls of the host,

(39)

causing human tissues to form abscesses. These eruptions break open, and thousands of eggs per day are passed in solid and liquid wastes to the outside. They hatch in water. The embryos swim about before burrowing into the snail's soft body. The snail-bound embryos develop fork-tailed larvae. They escape to the outside and remain very active in water until an unsuspecting final host — man — submerges some of his bare skin, usually his feet, in the infested water.

Other types of flukes enter man as he eats uncooked infested fish, water plants, crab, and crayfish meat. The fluke-to-be immediately burrows into the veins of man and seeks its adult hunting ground — the blood, liver, lungs, or intestines. And round and round the cycle goes, animal to animal.

With new drugs, which have been tested in Africa, the parasites were destroyed after a week of treatment. But drugs alone will not stop fluke infection. Reoccurrence may be prevented by educating the people, especially in tropical areas; by setting up sanitary measures for human waste disposal; and by eliminating the snail.

Working at Cornell University, Dr. B. H. Kean, a specialist in tropical medicine, and Dr. Edward I. Goldsmith, a surgeon, devised a method to remove flukes from man. They set up a system of internal tubes to pipe blood from the vein entering the patient's liver. A pump sends the blood through a filter that traps the ½-inch-long flukes but allows the whole blood to pass through. The blood is returned to a vein in the patient's leg. Drugs are used to make the worms scurry out of the main bloodstream and into the filter traps.

The first twenty trials of this system in Brazil resulted in the capture of 1,688 worms — all from one patient! Data on other cases

HEAD

NEW PROGLOTTIDS

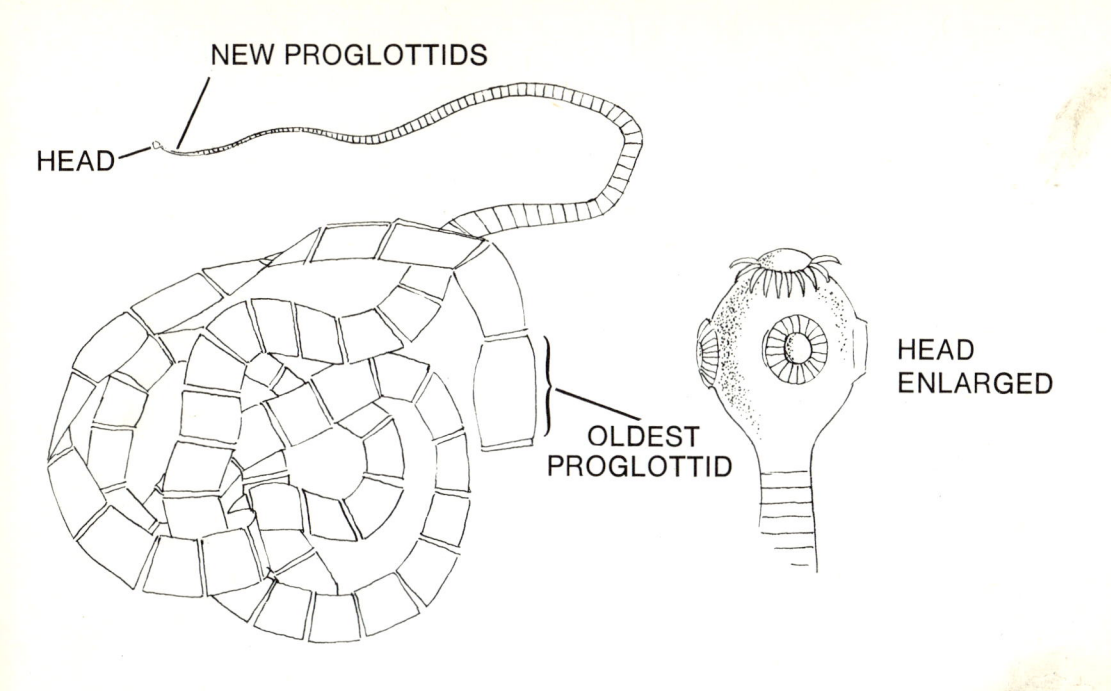

OLDEST PROGLOTTID

HEAD ENLARGED

indicated that the excretion of worm ova was reduced to very few or none. Thus, the filter system was a surgical success.

The longest of all the Platyhelminthes is the tapeworm, which belongs to the class Cestoidea. About fifteen hundred species invade the intestines of all invertebrates. But six very selective species choose to inhabit man.

A tapeworm consists of a tiny knoblike, quite efficient head, or "mother segment," that attaches to its host with suckers, hooks, or both. Behind the head, or *scolex*, are segments called *proglottids*. These oblong, flat, paper-thin units are self-supporting in all their life processes. They use the neighbor segment only for attachment. New proglottids are formed directly behind the head, and the older, larger ones are pushed downward through the host's intestines. Eventually, heavy with fertile eggs, they break off and are excreted with solid wastes.

Tapeworms 80 feet long have been found in man. They wave about in the intestines absorbing free, predigested liquid nourishment with their up to four thousand proglottids. One amazing fact about these worms is that they are not digested by the digestive processes going on all about them. They remain intact as they swim in the digestive juices of the host.

The tapeworm is not only the longest of its kind; it is also the laziest. Each segment simply absorbs, through the body wall, food that the host has worked to digest. The worm is not even equipped with a mouth or a digestive tube. There are no sense organs other than sensory cells along the entire body wall. The worm, as it enjoys its immunity to the host's process of digestion, is blessed with the ability to eat without being eaten!

Even in the reproductive process, the tapeworm does not have to exert itself looking for a mate. Both sexes are present in each proglottid. The mature end of the body chain provides sufficient numbers of fertilized eggs to pass to the outside for the creation of worms-to-be. The eggs are picked up in untreated human wastes by intermediate hosts, which include cattle, fish, dogs, rodents, and other vertebrates.

The beef tapeworm commonly infests man in the United States. After the eggs escape in human wastes, they are picked up by the cow in the field. The embryos escape from the eggs in the cow's intestine, penetrate the blood vessels, and migrate into the muscles. They form cysts, a kind of resting stage for things to come. Since only thorough heating will kill the encysted worm, man is infested when he eats poorly cooked beef. His intestines give the cysts a dark, warm, moist place for a new scolex to emerge, attach itself, and produce proglottids. Other tapeworms, such as those found in pork, follow a similar cycle. Sometimes the worms involve two intermediate hosts.

Treatment for tapeworm infestation is not simple. The patient must do without solid food for twenty-four hours and undergo a thorough cleansing of the intestinal tract, using prescribed salts. An *antihelmintic*, or worm medicine, must then be used and is often administered through a special tube. It goes directly into the area where the small intestine joins the stomach. In a medical laboratory, technicians examine the patient's body wastes and hunt for the worm's head, or scolex.

In freeing man from his guest, it is important to know that the scolex has been evicted. Otherwise, new growth will take place from the same mother segment. Laboratory examination of the waste materials is the only way of knowing about the success of the treatment. Most people believe that the worst thing about tapeworm infestation in man is that he is robbed of his food. But the real problem is the damage done to man's intestinal wall by the worm's hooks and suckers, and by the toxic substances released by the pest.

The tapeworm problem can, however, be solved easily for man. He should eat only thoroughly cooked meat.

In an issue of *Public Health Reports,* a publication of the United States Department of Health, Education, and Welfare, government inspectors were said to have found sixteen thousand tapeworm cases among twenty-eight million cattle in 1967. About four-fifths of the slaughtered animals are inspected by looking only at the heart and jaw muscles. These are the easiest to examine. But other muscles may contains cysts, too. Therefore, even meat marked "inspected" may cause tapeworm infestation.

Certain precautions have been effective in limiting tapeworm problems in man. Sanitary disposal of wastes, government inspection of meat, and public information materials describing the health hazard have helped. But each of us must help, too. For instance, should you really insist, "Make mine *rare*, please!"

6. The Pesty Roundworms

When swarms of insects flit about you, you know it. You swat and spray and swish them away. On the other hand, you have been surrounded, maybe invaded, by roundworms, yet you did nothing. You did not even know they were near.

The twelve thousand known species of this group, belonging to the class Nematoda, differ from flatworms in shape and in the absence of suckers, cilia, or other attachment equipment. They have a complete digestive tract and separate sexes. They are bilaterally symmetrical and have two body openings — a mouth at the anterior end and a posterior anus for excretion.

Nematodes are said to be among the most abundant and widespread of all creatures. If all matter except roundworm remains were swept off the earth by magic, it would be possible to determine, by observing the worm material, where the lakes, rivers, and oceans once were located. By using worm remains, we could recognize where certain types of soils once existed, where cities and many humans may have been, and even the kinds and numbers of large plants, such as trees, that lived on the planet. Each organism and each type of environment would leave its own special type of nematode "identification tag."

A sample of garden soil may yield millions of roundworms that look like active white threads whipping about in the soil moisture. After you save drainage from house plants, put some in a clear glass dish. Under a bright light, you may be able to see the worm-whips with a magnifying glass. If you have a small microscope, place some of the water on a slide and watch for transparent threads to lash out in the light. The movement is not so much for the purpose of locomotion as it is for food-searching. You will find that the worm stays in view for some time. When the body wall is in touch with a solid, such as soil particles or the walls of an animal's intestine, friction plus the movement will tend to push the worm along.

If you have no luck with your plant drainage material, bring in a container of rich garden soil or a sample of salt water, sand, and mud. Keep it in a warm place, and be patient.

In the meantime, you might try to find some vinegar eels in

Trichinella spiralis "tours" through man.

HUMAN INTESTINE

CYST

CYSTS IN MUSCLE

LARVAE IN BLOOD VESSEL

mother of vinegar, the cloudy substance found in the bottom of a vinegar bottle that has been tucked away on the back of the pantry shelf for a long time. The "mother" is a fungus growth on which the worms feed. But most of the bottled products have been pasteurized to prevent worm growth. Try to get some bulk vinegar. One type of "eel" was discovered in Germany whipping about in the felt mats on which beer drinkers placed their moist mugs. Live and preserved specimens may also be purchased from a biological supply house.

The nematodes have smooth, non-segmented, slender, cylindrical bodies. They are rounded at the anterior end and pointed at the posterior end. There are fourteen species of worms known to cause man some medical problems — the intestinal parasites and the filaria worms, the blood and lymph parasites.

Trichinosis is a common disease in the United States. About 19 percent of the population is infested, and many cases are undiagnosed. *Trichinella spiralis* enters man in the form of cysts present in poorly cooked pork. The adult male and female become mature within two or three days. At that time the female punctures the lining of the small intestine and deposits her approximately fifteen hundred larvae so that they can invade the blood and lymph vessels. The human host may complain of extreme muscular pains, diarrhea, nausea, and fever. Incorrect diagnosis during this phase can be serious, because this is the time for medical action. Drugs can be given to cleanse the host's intestines of the young worms.

Frequently, intestinal flu or food poisoning get credit for the symptoms. If the patient is not treated, the worms finally twist themselves into spirals and form a case, or cyst, about themselves, locating in any muscles throughout the entire body.

Trichinella spiralis is the villain in the sad story of "Farmer

Farmer Brown studies his runt.

Brown and His Little Pig." The farmer noticed that one of his pigs was a runt — not as large as the other animals. So, Farmer Brown decided to segregate it and to feed it the best of diets to fatten it up. However, the little pig would not grow, or get fatter or even livelier. So, off to the slaughterhouse went the runt. It returned to the farm in neat packages ready for the freezer — pork chops, hams, bacon, pigs' feet, sausage, and pork butts. Then came the holidays. The neighbors were delighted with Farmer Brown's generosity with his pork, until things began to happen.

Families in the community began to complain of fevers, body aches, puffy eyes, diarrhea, and nausea — not all families, but many. The village was so badly stricken that health investigators moved in and finally pointed an accusing finger at the runt's remains. Examination of sections of the meat left in the Brown freezer revealed hundreds of thousands of cysts per ounce of pork, just waiting to dissolve in someone's intestine and to cause more trouble.

Today most farmers destroy any pig that does not grow normally. Farmers also try to keep their land rat-free, since three-quarters of all these rodents have trichina worms. Pigs eat rats and acquire the infestation.

The trichina worm can be kept under control. When garbage is fed to pigs, it should be cooked first because of the raw pork scraps that are frequently included. People should discourage the eating of raw sausage, rare pork, or poorly cooked meat mixtures of any variety. Frankfurters and hamburger often contain some pork. Inspection of meat is ineffective since the cysts are microscopic and only paper-thin shavings can be examined. You are the best control of all.

Hookworm (*Necator americanus*, "the American killer") is a very old human disease. As early as 1500 B.C., the Egyptians recorded their knowledge of parasitic roundworms, listing some remedies for ridding man of the pests. They recommended such things as goose fat, turpentine, and certain plant roots. Some of the cures proved worse than the disease. The Chinese, in A.D. 200, were aware of "the long worm," which their writings claim to have been 5 to 12 inches in length.

The infection is picked up when bare skin, usually on the feet, contacts soil where the larvae pierce the victim. They travel by the blood to the lungs, where they ascend through the bronchi and trachea, are swallowed and develop, in the small intestine, into adult worms less than one inch long. They attach themselves by a strong mouthpart sometimes equipped with teeth; they then suck blood, which they manage to keep from clotting by secreting a

INTESTINE

Life cycle of the common hookworm

juice around the place of attachment. Thus, their liquid diet is uninterrupted.

Heavy infestations of hookworm cause chronic anemia or a deficiency in both the quantity and quality of the blood. This results in general weakness and fatigue. Treatment by a doctor can usually completely rid the patient of the disease, but the best treatment is prevention.

Hookworms mate in the host's intestine, and thousands of fertile eggs pass out in waste materials. When unsanitary disposal of human wastes is found, hookworm disease abounds. General cleanliness and the wearing of sandals or shoes will help man to curb the spread of this common infection. Dogs, cats, foxes, sheep, and pigs also act as hosts to other hookworm parasites.

Ascaris lumbricoides is the giant of the nematodes. Some have been found more than 12 inches in length. The adults, both male

Intestinal roundworms of pig and man are similar in structure, but neither pig nor man usually infests each other with Ascaris worms.

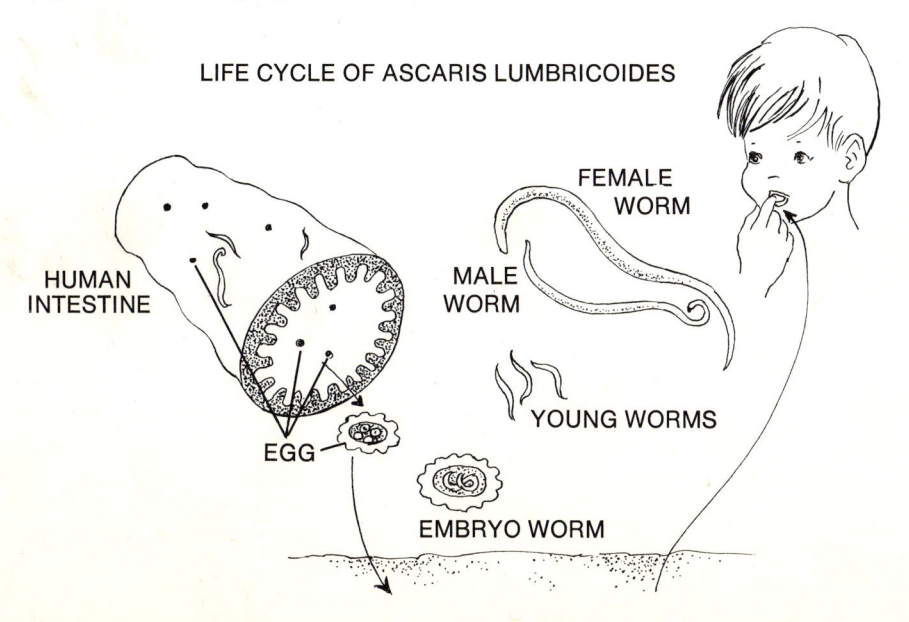

LIFE CYCLE OF ASCARIS LUMBRICOIDES

HUMAN INTESTINE

FEMALE WORM

MALE WORM

YOUNG WORMS

EGG

EMBRYO WORM

FILARIAL WORM— A THREADLIKE ROUNDWORM

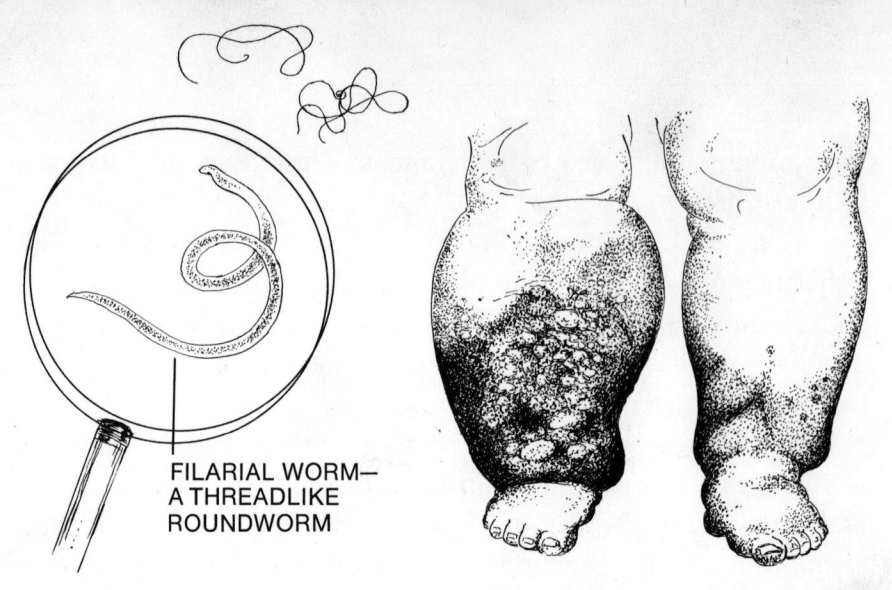

Elephantiasis, a disease in man, is caused by filarial worms.

and female, live attached to man's intestinal wall and share his semidigested meals. No intermediate host is necessary to complete the life cycle.

When numerous worms are present, the intestines of the host can be completely obstructed. At times, the parasites migrate to the nose or mouth or even penetrate the intestinal wall and invade other organs. This can kill the host.

Pinworms or seat worms are common human companions, especially during childhood. They enter the body as eggs, transferred from the hands to the mouth. Sometimes they get into the

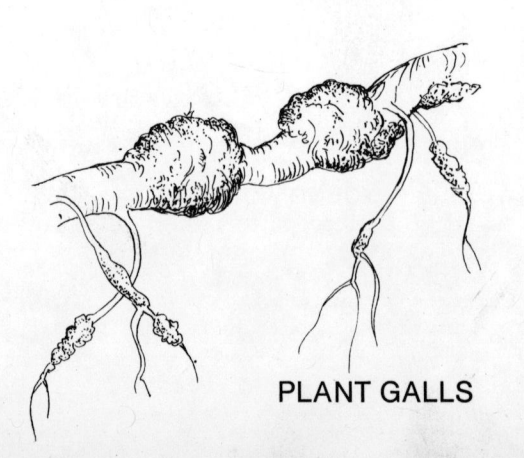

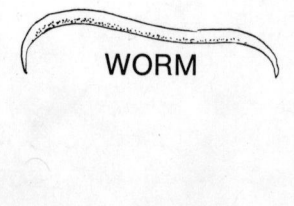

WORM

PLANT GALLS

body in contaminated food or water. The eggs hatch in the human intestine and the ½-inch male and female worms mate. The female migrates to the anus to deposit her eggs. The host suffers from an unbearable rectal itch during this time, which usually takes place each evening. It is only by extreme hand and fingernail cleanliness and daily changes of clothing and bed linen that control can be effective. Potent drugs will evict the worms within a week, but reinfestation is the big problem.

There are six filarial worms of medical importance in the tropical regions of the world. These threadlike roundworms invade human muscles and block the lymph channels, having used mosquitoes or flies as intermediate hosts. A disease known as elephantiasis, characterized by immense swelling in man's muscles, is the result of this worm's invasion.

It is interesting to note that nematodes live on plants as well as on animals. The sugar-beet roundworm limits itself to that plant, but root-knot nematodes have been found in more than one thousand plant varieties, including fruit and shade trees, shrubs, and weeds. You have probably seen swollen spots on roots. These are called galls and are made of scar tissue. Thus, the plant reacts to the burrowing larvae that are feeding on the root tissue. Try to find some galls or root-knots, cut them open and examine them with your magnifying glass. Control is difficult because of constant contamination of the surrounding soil.

Worm controls are man's problem, and it seems that sanitation and personal cleanliness may be the answer. Plants can act as worm controls, too. Most people know that the Venus flytrap is an animal-eating plant, but there is a nematode-trapping fungus that was discovered about 100 years ago. The plant works in the soil and has sticky sections like flypaper that catch the worms. Then the plant digests the animals. Another fungus plant produces adhesive ring traps that constrict when the worm touches them, making it impossible for the worm to escape.

The most interesting plant trap is on the fungus that has small spherical knobs protruding from short stalks. The knobs are coated with a poison secretion, making a worm trap that has been called a "lethal lollipop."

You may never have seen a roundworm, but they exist. They are pests, and they must be kept under control.

7. Earthworms—the Underground Wonders

Worms means earthworms to most people. As you know, many other related species exist and have been studied and classified. The anatomy of each member of the phylum Annelida is very complex. The general body features of the 8,500-member phylum include ringlike segments, or *somites*. They can be seen in the internal and external structure of the bilaterally symmetrical body. There are paired bristles, or *setae*, on each somite. The digestive tract has two openings, an anterior mouth and a posterior anus.

The organ systems include clearly defined equipment for every function found in man. There is a highly developed nervous system connected to two main nerve centers, or "brains." Specific organs are assigned to the work of circulation, digestion, excretion, respiration, and reproduction, with variations for each of the six classes of annelids.

The phylum includes earthworms, sandworms, tube worms, leeches, and many small marine worms. Let's pry into the life of the not-so-lowly *Lumbricus terrestris*, the common earthworm.

EARTHWORM

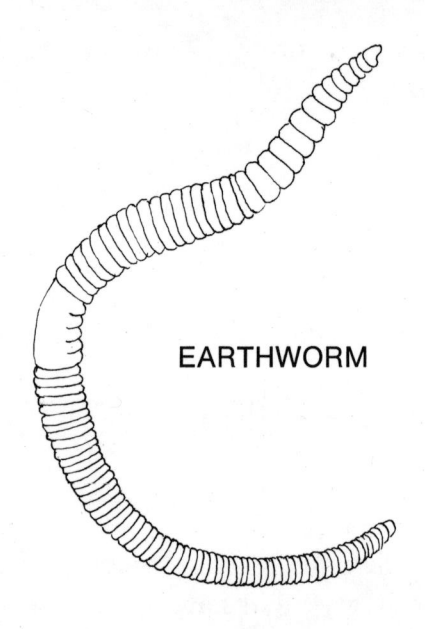

This animal is known by a variety of names, such as angleworm, dew worm, night crawler, fishworm, and rainworm.

You can discover many things about the earthworm's exterior by using your magnifying glass and observing any one of the more than one thousand species that have been identified thus far. Look for a patch of heavy, rich, moist soil, and turn over the dirt to about 18 inches. Mild weather will bring the worms closer to the sod, but they have been seen mating above ground in forty-degree temperatures. By day, it is difficult to see the opening to an underground tunnel because the creatures cover the entrance with seeds, twigs, or pebbles.

After a heavy spring rain, when worms are said to be "rained down," they are found above ground in great numbers. They were "evicted" from their burrows by flooding. Since the animals breathe through their body walls and live very comfortably submerged in water for long periods of time, it is incorrect to say that in waterlogged burrows they may be drowned. They are actually starved for oxygen during heavy rains because there is less oxygen dissolved in water than is found in the air. The worms' only hope for survival is to emerge to the open spaces, where they are faced with still another problem. Above ground they are frequently crushed or die as the rains cease and their skins dry because of evaporation. The wise worm keeps its burrow entrance plugged up so that flooding will not become one of life's hazards.

Night collecting is sometimes easier, since the worms come to the surface for feeding, mating, and migration. Although it has no eyes, the earthworm's photosensitive body causes it to be sensitive to white light but leaves it insensitive to red. So, cover your flashlight with red cellophane and you may have luck. Later, you may

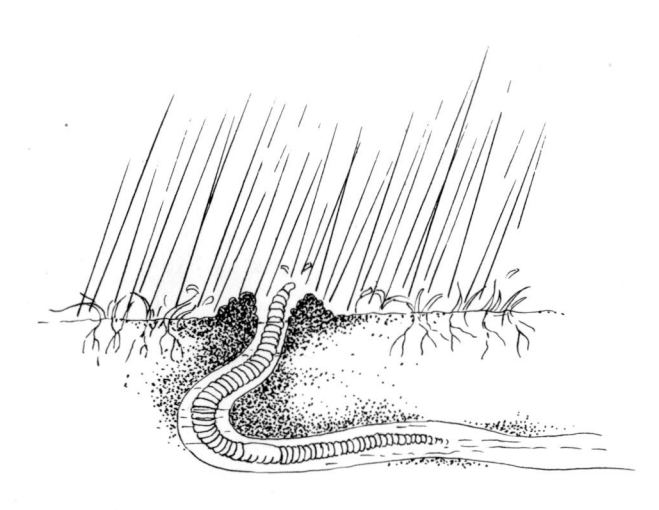

Flooding evicts oxygen-starved earthworm.

"Fiddling" for worms is fun.

experiment with blue light and watch with amazement as the eyeless animal withdraws.

The earthworm has no ears. But tapping the ground's surface causes the vibration-sensitive animal to seek peace and quiet above ground. Experienced worm-trappers use battery-operated vibrators, but "fiddling" is more fun. This means that you must drive a stake about 10 inches into the soil, and rub the side of a thin board across the top of the stake until the worms come up. One fisherman reported that after using this technique, his bait-to-be appeared in a radius of 25 to 30 feet around the fiddling point. If no earthworms appear, you may have chosen a poor hunting ground or the wrong season.

The best storehouse for an earthworm collection is a tight wooden box or a stone crock containing 8 inches of rich, moist earth and a 2-inch topping of pieces of plants. Keep it in a cool, dark place. Occasionally, put in bits of chopped raw beef suet, crumbled hard-boiled egg, or fine bread crumbs. Sprinkle these tidbits on the soil under the plant layer.

Commercial earthworm farms are located in California and in some southern states. The animals are fed a balanced diet. When they are about three months old, they are packed in soil and moist peat moss and shipped to farmers everywhere. The best commercial worms have been "scoured," a process known for hundreds of years. Scouring is done by packing sphagnum moss into a stoneware crock or tight wooden box. The moss is kept well moistened, but excess water is wrung out before use. The worms are stored in the moss for two to four days and kept in a cool place. The bait becomes almost transparent, tough, and very lively. The animals will live longer on the fisherman's hook and will lure more fish if they have been scoured.

Farms also sell worm eggs by the quart to farmers for the production of "living plows." They are in ¼-inch-long packets, and each can produce up to a dozen worms under favorable conditions. So, the farmer who seeds his soil may be planting worms, not vegetables.

You make many observations if you work with live worms, but there are some things you will want to know before you start. The worm's body is covered with a thin transparent *cuticle*, which is lubricated and kept soft by mucus produced by the *epidermis*, or outer body layer. There are external openings, such as the *dorsal pores*, or openings to passageways from the body cavity to the exterior. These are found middorsally in the furrow on each somite from number VIII or IX to the anal end. Fluid passes through

Does the earthworm look like a "simple" animal?

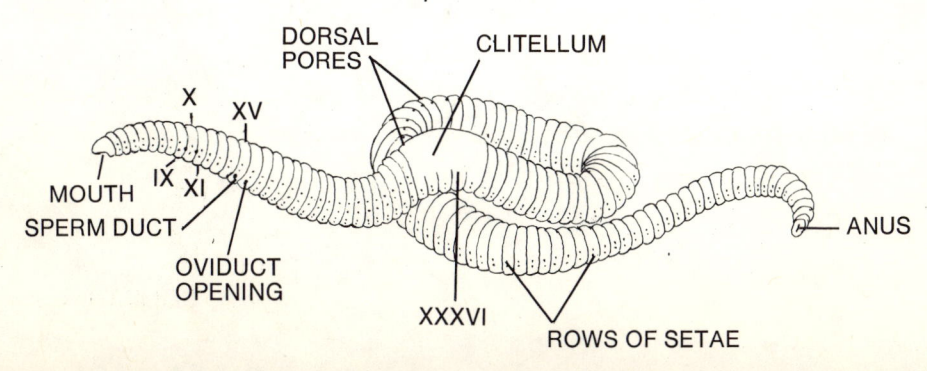

these pores from the worm's interior. This, plus the mucus, keeps the cuticle in condition for respiration. Of course, environmental moisture is important, too, and you will find that hard, dry, caked soil is poor worm-land.

An excretory pore is found on each side of the somites, except on the first three and the last. *Seminal receptacles* are located in the grooves between somites IX and X and somites X and XI; the two *oviduct openings* are located on somite XIV; and the fleshy ridged *spermaduct opening* is located on somite XV. The worm is hermaphroditic, that is, each individual has both male and female sex organs.

There is a swollen area, the *clitellum* (Latin, meaning "pack saddle"), which covers several segments and secretes cocoon-forming material to contain the eggs and sperm. Earthworms reproduce throughout much of the year, but are more active in warm, moist environments. Mating takes place at night as the worms stretch out from their burrows and lie with the ventral anterior surfaces together, their posterior ends pointing in opposite directions. Special setae near the mating area hold the animals together as each worm secretes a slime tube about itself from somite IX to XXXVI. The three-hour period of exchanging sperm ends as the worms

Earthworms mate, or exchange sex cells, above the ground.

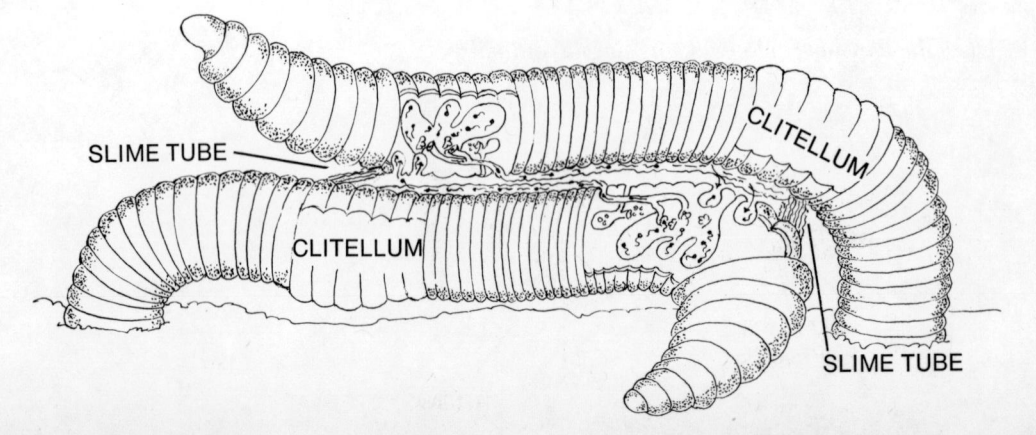

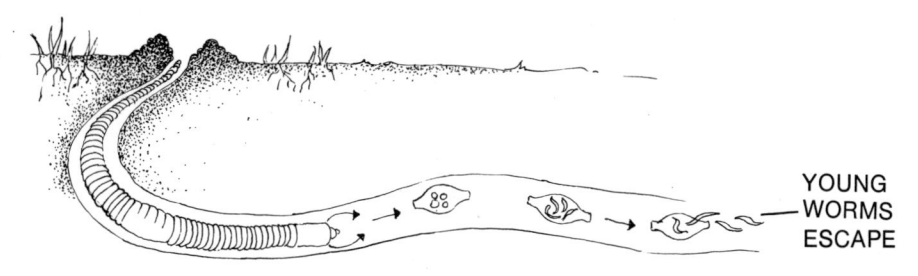

Cocoons with fertile eggs stay in moist soil until the young worms develop and escape.

separate. Fertilization takes place in the cocoon. The worm withdraws from the cocoon, which seals at the ends and stays in moist soil for two to three weeks. By then, most of its two dozen eggs develop into baby worms. The newborn wiggle out of the case to eat their first meal — soil! This is quite a complicated process for animals that have been called simple and lowly.

Now you are ready to launch your own investigation. The Earthworm Observation Sheet on pages 62 and 63 is just a starter for you. Develop your own method of recording as you become more experienced and as you create your own problems. Be sure to refer to the Worm Observation Sheet used in Chapter 2. You will find that with your additional knowledge, this guide can be more meaningful. For example, vary the color of the light under Section B on Movements, and record both the color used and the worm's reaction.

The internal anatomy of the earthworm is structured for its way of life. Its natural diet consists of decaying plants, grasses, seeds, and small animal material. This worm food is often mixed in the soil of the earth, so the animal actually lives as a chemist, a borer, and a living plow. The soil pushed into the mouth by the *prostomium,* or lip flap, passes into the *pharynx* inside somites IV and V, on to the *esophagus* (VI–XIV), then to the *crop* (XV–XVII) for storage, and on to the sand-filled *gizzard* (XVII and

Earthworm Observation Sheet

Problem: What external features are present on an earthworm's body and how are they related to the animal's way of life?

Note: Choose about five active worms and place them in a box with moist rich soil. Work with one worm at a time using a separate tray and a magnifying glass.

A. *Comparison of ventral and dorsal surface:*

	Number of species observed				
	1	2	3	4	5
1. Dorsal darker than ventral (yes, no)					
2. Ventral flatter than dorsal					
3. Dorsal and ventral, same color					
4. Dorsal and ventral, same shape					

B. *Somite Study:*

1. Number of somites
2. The clitellum covers which somites? (Give numbers of those covered: from ____ to ____)
   Note: the *clitellum* is a glandular swelling toward anterior.
3. The clitellum covers how many somites?

C. *Setae Study:*

1. On what segments are there none?
2. On what surface are they? (D–dorsal; V–ventral; L–lateral; or a combination—DL, VL)
3. Run a worm through your fingers. Do you feel the setae? Of what use are they to the animal? Make some notes on your ideas.

D. *Body Openings:*
Using the information given to you in the text, find the opening and record the number of the segment on which it is located.

1. Seminal receptacle openings
2. Oviduct openings
3. Sperma duct openings

E. Other notes, comments, and questions:

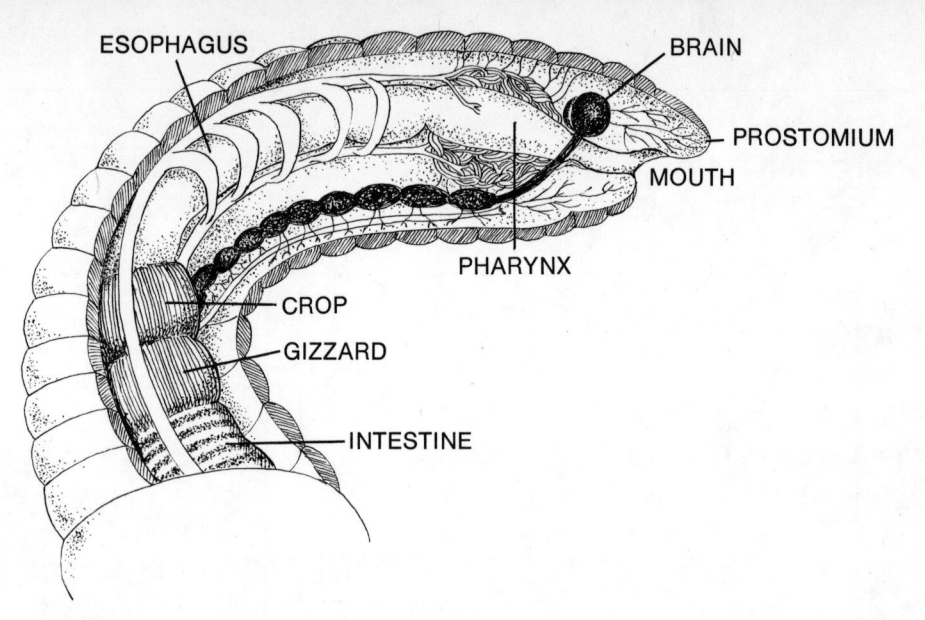

A worm's meal follows a set path to the intestine, then on to the anus with undigested waste materials.

XVIII) for grinding. The intestine absorbs digested material passed from the gizzard, and wastes are passed out through the vertical anal opening in the form of castings. These familiar piles of excretory material around the openings of an underground habitat mean that a worm has digested a meal, that some formerly subsurface soil has been transported through the worm's gut to the outside, and that the material has been chemically transformed into nature's ideal plant food.

In 1882, Charles Darwin wrote *The Formation of Vegetable Mould Through the Action of Worms, with Observations of Their Habits*. He said, "The plough is one of the most ancient and most valuable of man's inventions, but long before it existed, the land was in fact regularly ploughed and still continues to be thus ploughed by earthworms."

Darwin claimed that the whole body of topsoil has passed,

and will pass again and again, through the bodies of worms. He collected and weighed worm castings, estimating that between 7½ and 18 tons of material are brought each year to the surface in each acre of land. Today, this "worm manure" is sifted from compost culture beds, and 100-pound sacks are sold to florists for fertilizer.

You may have heard that if you cut an earthworm in half, you will be able to regenerate new worms from each part. This is not true. The area of the cut determines the success of the new growth, and even then, it varies with the species. Most earthworms can replace four or five segments at the anterior end, which is more successful in making the adjustment, but a cut between segments XI and XXXVI will kill the entire worm. Do not let the wiggles fool you. Waves of peristaltic movement (similar to waves passing down a rope) pass down through the half worms, and each segment is stimulated by the motion of the neighbor segment.

Many interesting facts have been discovered concerning the earthworm's behavior. The animal has been trained to go right or left at a fork in a T-shaped glass tube used as a maze. One arm of the apparatus led into a jar of moist earth and moss, and the other, lined with sandpaper, was equipped with electrodes that gave the worm weak electric shocks. After about one hundred trials, the worm learned to choose the moss trail. Even the best time for learning has been determined. It seems that from eight in the evening until midnight, the maze can be conquered with fewer trials. This relates to the animal's nocturnal habits. Learning takes place even if the central nervous system of the first six segments has been removed. Thus, each segment is capable of independent learning. The credit is not given to the worm's intelligence, but to habit formation.

When worms are not eating, they are "sleeping." Worm-sleep for this eyeless creature simply means sustained inactive periods underground. Some worms have been observed following a forty-minute schedule of eating and resting. It appears that these animals have a built-in biological clock that is controlled by the nervous system in the esophagus.

Darwin knew that earthworms can distinguish between foods, and the selection of materials seems to be related to taste. Scientists have boiled pine needles to remove their original flavor. Then they coated them with gelatin mixed with powdered leaves. The earthworms have shown a preference for flavors in this order: beech, maple, oak, and horse chestnut.

The next time you see an earthworm, you might think of it as one of the underground wonders of the world!

8. More Worms With Rings

Worms with rings — the annelids — have no housing problems. They inhabit manure or compost piles, damp lakeshores, decaying seaweed beds, fresh or stagnant water, ocean shores, mountain soils, and coral reefs. Most of them are vegetarians and enjoy dead or decaying plant material. They exhibit many of the physical features of the earthworm, and their life requirements are similar. Other interesting annelids include the sandworms and tube worms, the exciting palolo and fireworms, and the leeches.

The sandworm, or clam worm, *Nereis virens,* is a saltwater, low-tide-line inhabitant. It has a definite head with tentacles and two pairs of eyes. It is named after the Greek sea nymphs, the Nereids, although there is no physical resemblance. Careless handling near the anterior end can result in a painful nip. The worm protrudes its pharynx and chops down on a fisherman's finger with two horny-toothed jaws. It then tries to pull the pharynx back into place, and the victim's skin goes with it. If the captive is a small animal on which the sandworm thrives, it can be torn to shreds by this process.

Stiff bristles give support to the *parapodia* (from the Greek

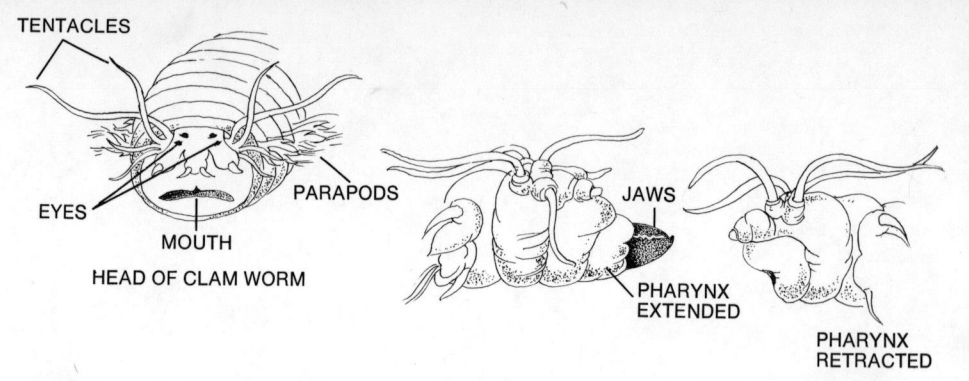

The anterior, or head, region of the sandworm, or clam worm, can be a busy area, and one of danger for the careless fisherman.

para, meaning "beside," and *podos,* meaning "foot"), which are used for creeping and swimming. They have given it the titles "paddle-footed" clam worm and rag worm. When lifted out of water, the oarlike extensions droop like old rags. The worm hides by day in a temporary burrow with its head showing topside. At night it roams about over the sand or swims by side-to-side body wriggling. This active night life is made possible by the parapodia, which connect directly with the body muscles in each segment.

Most sandworms have reddish brown bodies, which vary in length according to the species. They range from a few inches to 18 on the New England coast and to 3 feet on the Pacific shoreline. These Pacific "sea serpents" have been mistaken for snakes because of their size.

The sexes are separate. During reproduction, changes take place in the worm's physical appearance and in its activities. The eyes enlarge, the paddles stretch out and flatten, and the body color darkens. The male spirals through this period like a ballet dancer, while the female, whose posterior is becoming heavy with reddish eggs, joins him later in the dance. The male spins faster, and finally both sexes shed their sex cells, which sometimes explode through the torn body wall.

Fertilization takes place in the water. The young appear first as swimming larvae propelled by cilia, but later they changed to new worms. During this time, bathers will frequently leave the shore waters because "it is full of worms." And they are correct. Sometimes the high-tide line on shore is marked by masses of worms that failed to survive.

Sandworms live in temporary tubes in the sand, but a related species — called "the wanderers" — seldom stops long enough in one place to require an underground apartment. The scale worms carry armorlike shields on their dorsal side, made up of from twelve to fifteen overlapping pairs of plates. *Aphrodite aculeata* is called "sea mouse" because its back plates are covered with long hairs that are gray through the middle and iridescent green and gold on each side. Aphrodite, when an adult, is about 7 inches long and 3 inches in width, so it is really an oversize super-mouse if compared to its animal namesake.

Some annelids maintain one residence for life. These are the tube-dwellers. The tubes vary with the species, but are built to last. Some are constructed of sand and bits of shell, which are plastered with mucus to produce a smooth lining. One type of tube acts as a "worm-trailer." It is built of cemented sand grains. The mobile home is carried around with the worm called *pectinaria*. *Spirorbis*, another tube-dweller, secretes a hollow spiral tube of strong calcerous material and anchors it to rocks or seaweed. Some tube worms bore into rocks or shells to set up living quarters instead of constructing a home.

The problems of the tube-dwellers have been solved by the parchment worm. The creature builds a permanent U-shaped tube, about ¾ inch in diameter, made of a secretion from the worm's body that forms a parchmentlike material. The tube-home is at

PEACOCK WORM

**TUBE WORM
(SPIRORBIS LAEVIS)**

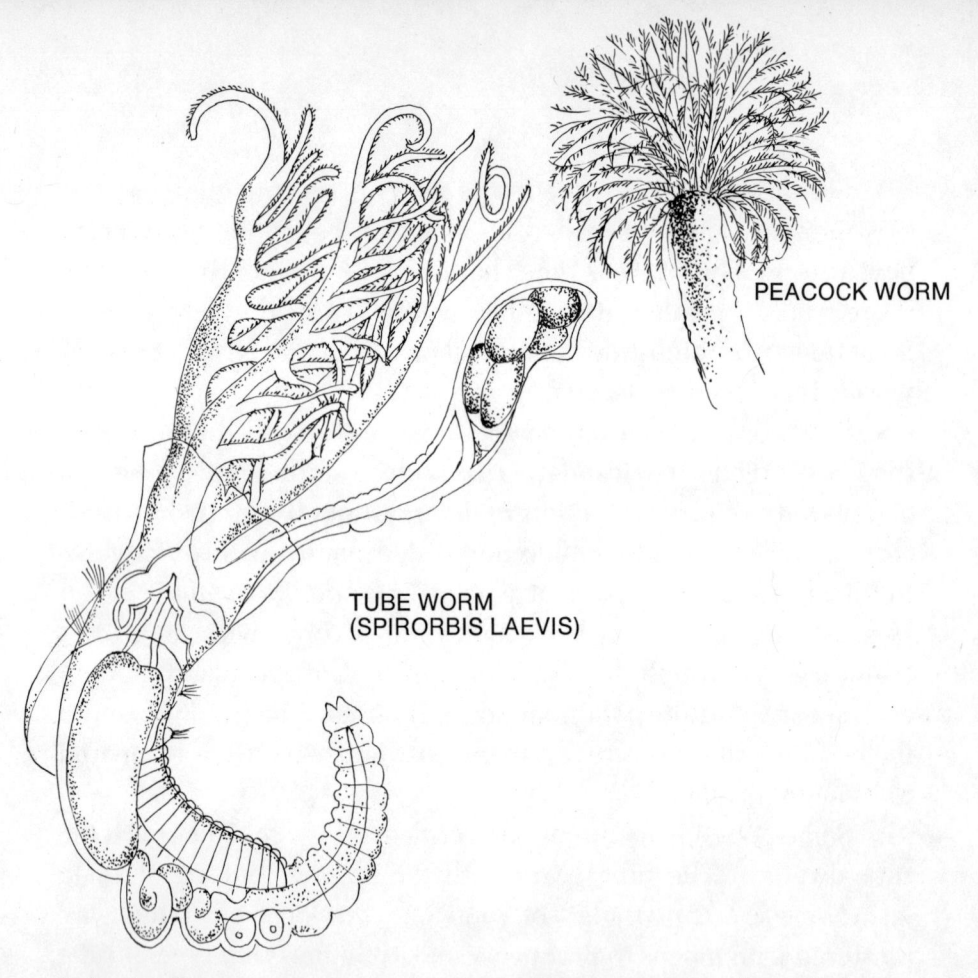

Tube-dwelling worms have no housing problems, except to keep fresh currents of water and food moving into their apartments.

least three to four times as long as its tenant, who stays most of the time in the bottom horizontal zone. The animal is equipped with three pairs of special parapodia in the middle of the body. They act as fans to keep a current of water and nourishment flowing through the tube. Food particles are trapped in mucus from the worm and are rolled into balls. When the creature is hungry, these pre-prepared meals are engulfed by its wide funnel-like mouth.

When waste materials are ready for excretion, the worm moves backward until it deposits a casting above ground. If sand or debris clog the tube, the material is taken in, moved through the digestive tract to absorb any nourishment, and then another casting of wastes is formed.

This worm also takes in roomers, tiny crabs which live with their landlord as commensals. *Commensalism* is a form of existence in which animals of different species live together, with one receiving some benefit as the other remains unharmed. It is suspected that the roomer setup is desired by both animals and comes about not just by accident. The crabs inhabit the area at the anterior end of the tube worm, so it may be that the crab benefits shelterwise while the worm relies on the water currents whipped up by the crabs. Crabs are scavengers, so it is possible that the worm's housekeeping chores are shared.

The luminescent flashing of worms may have been responsible for the lights that Columbus reported the night before he sighted land. Some tube-dwellers give out azure blue to greenish light, so bright one can read printed letters by it. The luminescence is sometimes related to irritation, as when the creatures are suddenly disturbed. The flashes may send an enemy about its business. The light-giving process is also related to the reproductive period when mates must be attracted.

In Bermuda and Florida, "fireworms" swarm on the second, third, and fourth days after a full moon. They come to the surface to spawn exactly fifty-five minutes after sunset. The female starts the "lamp-lighting" as she circles about, basking in her own green-

PALOLO WORM

ish glow. The male then sets out to meet her, but his lights are more like the dots and dashes of a telegraph key. He is actually sending out mating messages. When the sexes meet, they rotate together, scattering eggs and sperm — then the lights go out.

The natives of the Samoa and Fiji islands celebrate what could be called "worm holidays." For weeks before the big days, the villagers gather buckets, baskets, pails, flashlights, lanterns, and nets.

(72)

On midnight of the eighth day after the full November moon, when the sky is dark, and for three consecutive days, the islanders sit on the banks by the sea, waiting for a signal from the "reef-watchers." These natives, armed with lanterns, move back and forth across the waters, peering into the sea for signs of the big moment.

The cry *"Va sav le palolo"* — the palolo comes — is heard, and the people join the swarming worms in the water. They dip up thousands of the species *Eunice viridis,* an annelid that produces during its reproductive cycle 16-inch-long segments that are loaded with eggs and sperm. Each worm casts off in one unit the rear section of its body, which has developed an eyespot. This may help in guiding the castoffs toward the surface. The parent worms remain below in coral-reef holes and begin to develop new posterior reproductive sections for the next year.

The palolo worm segments are considered to be the choicest food of the year. Even while the islanders are collecting them, they are stuffing handfuls into their mouths. Later, the worms are wrapped in leaves and broiled on charcoal made from coconut shells. The name *"viridis"* means green. After cooking, the meal looks like spinach. Ten-pound worm bundles, each with a leaf "crust," are baked until dry, then hung up to be recooked in coconut cream sometime later.

If man can feed on worms, it seems fair to expect worms to do likewise to man. The annelid known as the leech likes nothing better than human blood. The animal has a long, or oval, segmented, very flexible body that lacks setae or appendages. It easily contracts or expands the body wall as it roams on moist land or swims in freshwater lakes and streams. Its grabbing and hanging-on power is supplied by two suckers located at each body ex-

LEECHES

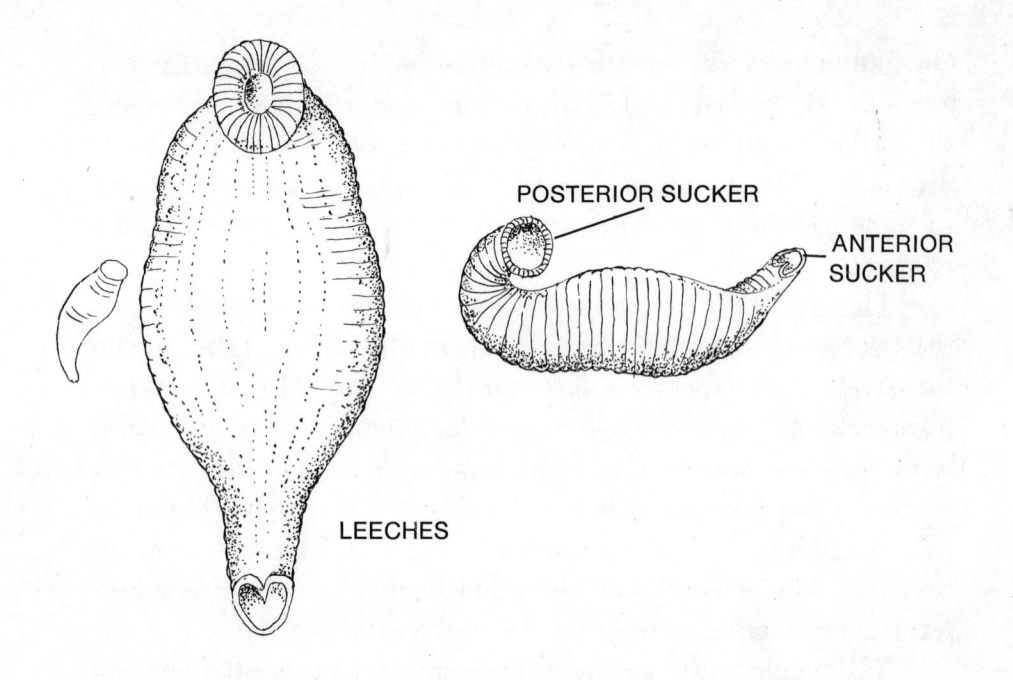

tremity. The large posterior apparatus is responsible for locomotion and attachment, while the smaller anterior sucker contains the mouth. The animal attaches itself to its victim, pierces the skin with tiny teeth, and makes a painless Y-shaped wound. It sucks blood until its digestive tract, plus the side storage pouches, can hold no more. It then drops off its victim and spends days digesting the meal.

Some land leeches take in ten times their own weight in blood before quitting. Imagine how big your meals would be if you could eat ten times your own weight before quitting! Not surprisingly, the animal requires only one or two meals a year.

The leech secretes a substance, *hirudin*, that prevents the blood from coagulating, or clotting. That explains why the wound bleeds for almost ten minutes after the leech departs. Experiments have shown that iodine does not sting when applied to the skin puncture. It is suspected that the animal secretes some form of anesthesia on the victim's skin.

The medicinal leech, *Hirudo medicinalis,* has been used since ancient times, when doctors who practiced "leechcraft" used the worms for bloodletting. This was supposed to drain "bad blood" from the body. It was possible in the old days to rent a leech from a pharmacy, which kept the animals in jars waiting for business. A leech, when applied to a bruise, would hang on until it sucked all the blood it could hold. Then it dropped off and was returned to the pharmacy. Unfortunately, blood diseases were easily transmitted through the "rent-a-leech" business.

Whether worms are flat or round or segmented, these creatures hold an impressive position on the ladder of nature.

The leech has teeth which cause victim to bleed from a Y-shaped wound.

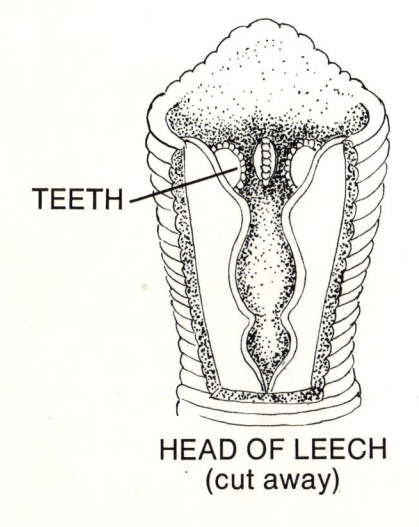

TEETH

HEAD OF LEECH
(cut away)

Y-SHAPED WOUND

9. Worms and Man

Man may consider worms as either helpful or harmful. Sometimes they are just annoying. In deciding to encourage or discourage the worm population, man must make some decisions that, if unwise, can have negative effects on generations of future living things.

"Earth Day" — April 22, 1970 — was dedicated to making individuals aware of the need for a clean, healthy environment in which to live. Pollution of air, water, and land is one of the key issues for society today. Only man, the thinking animal, can solve this problem.

Industrial and automobile wastes have been accused of fouling the envelope of air around the earth. But pesticides and herbicides must share the blame. Rain and snow act as scavengers to cleanse the air, but the pollutants are then deposited in the ocean and on the land. As animals feed on living things that have been produced and nourished in pollution, they actually concentrate these poisons in their bodies. The brains of dead and dying robins have been found to contain three to twenty times more DDT than the leaves that made up the diet of the earthworms that, in turn, nourished the robins. Fortunately, the sales of potent insecticides that may kill or control pests have been curtailed. It has been

found that these chemicals often harmed man as well as the parasites.

Beautiful beaches along the coasts of the United States are turning into hazardous sewers because of man's carelessness. Oil pollution from industrial and shipping spillage is killing, bit by bit, the wildlife that has been America's heritage. Not only the shorebirds and the fish are affected, but the sand dwellers — the worms — are rapidly decreasing in number. Or else they are adjusting to the filth and becoming poisoned bait for their predators.

Worms have a rightful place in the balance of nature. If man comes in contact with parasitic worms that can cause him physical harm, he must use precautions to prevent personal injury. He can avoid tapeworm and roundworm infestations by refusing to eat poorly cooked meats. In hookworm country, he should not go barefoot. Cleanliness and sanitary waste disposal can eliminate many of the worm hazards for man.

Sometimes worm castings can be unwelcome additions above ground. National golf tournaments have been halted while a player asks permission to have a worm-hill removed before he putts. Again, man has a choice — to poison-spray the entire course or to stamp down the occasional obstacle.

Fertilizers can be harmless. One liquid product, the color of tea, is odorless. Yet flies, ants, sow bugs, and many garden pests find it repulsive. It is made from earthworm castings and contains elements of concentrated plant food. But it takes ten thousand worms about one hour to produce the ingredients for just one gallon of this product, so the demand exceeds the worms' ability to supply.

Man has demonstrated fantastic success as a crossbreeder of animals and has developed new varieties to fit his needs. He has experimented in the "worms-made-to-order" business and has produced hardworking worms for the farmer; giant or pudgy worms, rich in protein, for the chicken farmer; and little worms with a big wiggle for the fish hatcheries.

Man must use his intelligence to maintain the natural world's balance in a healthy, unpolluted package of land, air, and water — his own balance is in danger. Worms are just one rich segment of the earth's population that is dependent upon man.

Index

Acoels, 19
Annelids, 5, 55-66, 67-75. *See also*
   specific animals, i.e., Earthworms
      diet, 67
      general anatomy, 55
      habitats, 67
      organ systems, 55
      species, 55
      subdivisions, 5
Aphrodite aculeata, 69
Apple "worm," 57
Archannelida, 5
Aristotle, "ladder of nature," 4
      interest in worms, 35
Ascaris lumbricoides, 51
Aschelminthes, 5
Asymmetrical symmetry, 17

Beach pollution, 77
Bermuda "fireworms," 71
Bilateral symmetry, 17-18
Binomial system, 4
Blood fluke, 37-40
      controls, 40
      in Egyptian mummies, 38
      life cycle, 39-40
      Schistosomiasis, 37
      snails as host, 38
      surgical removal, 40-41
Bloodletting, 75
Bristles on worms, 55, 68

Castings, 64, 71, 80
Caterpillars, 7
Cestoidea (Cestoda), 5, 35, 41
Chlorophyll in worms, 19
Clam worm, 67-69
Classification, 4, 6
      by color, 19
      origin of names, 4
      process of classifying, 6, 19
      by size, 20
      subdivisions, 4
      by symmetry, 17-19
Clothes-moth larvae, 7
Codling moth larvae, 7
Collecting worms, 9
      earthworms, 9-10

indoor laboratory, 10
meat traps, 11
outdoor laboratory, 10
planaria, 11
roundworms, 14-15
sandworms, 15-16
tools, 9
Columbus and worm light, 71
Commensalism, 71
Convoluta, 19
Crabs and tube worms, 71

Darwin, Charles, and earthworms, 64-65, 67
Data-keeping, 10
DDT, effect on animals, 77

Earthworms, 5, 55-66
    behavior research, 65
    collecting, 9-10
    collecting by "fiddling," 58
    collecting at night, 57
    common names, 56
    earthworm farms, 58
    eggs, 59
    external anatomy, 59-60
    food preferences, 66
    internal anatomy, 61-64
    life cycle, 60
    observation of earthworms, 56
    scouring, 59
    sleep, 67
    used for food, 20
Elephantiasis, 53
Eunice viridis, 73

Fertilizer from worms, 80

Filarial worms, 50
Firefly larvae, 35-36
"Fireworms," 71
Flatworms, 4, 35-44
Flukes, 5, 35-41
    anatomy, 35-36
    bloodfluke, 37
    fish parasites, 36
    fluke disease, 40
    fluke hosts, 38
    fluke traps, 40-41
    importance of flukes, 36-37
    life cycle, 39
    Schistosomiasis, 37
    size, 20
    species, 39, 40
    subdivisions, 5
    "Swimmer's Itch," 37

Glowworm, 7
Gnathostomulida, 6
Goldsmith, Dr. Edward, 40

Hermaphrodite, 60
Hirudin in leech, 74
Hirudinea, 5
Hirudo medicinalis, 75
Hookworm, 49-51
    anatomy, 49
    ancient records of, 49
    control, 51
    disease caused by, 51
    life cycle, 51
    treatment for disease, 51

Inchworm, 7
Insecticide controls, 78

Kean, Dr. B. H., 40

Larvae, 7
    caterpillar, 7
    clothes moth, 7
    codling moth, 7
    inchworm, 7
    mealworm, 7
    ringworm, 7
    silkworm, 7
Leechcraft, 75
Leeches, 5, 73-75
    exterior anatomy, 73
    feeding process, 74
    Hirudo medicinalis, 75
Linne, Carl von (Linnaeus), 4
    binomial system, 4
Lumbricus terrestris, 55
Luminescent worms, 71

McConnell, James, 32
    planaria research, 32
Marine worms, 5
    new group, 6
Mealworm, 7
Medicinal leech, 75
Moth larvae, 7

Names of worms, 4-5
Nematoda, 5, 45-54. *See also* Roundworms
    Ascaris lumbricoides, 51
    Necator americanus, 49-51
    nematode plant traps, 53
    Trichinella spiralis, 47
Nereis virens, 67

Observation of worms, 25

Ocelli, 24
Oligochaeta, 5

Palolo worm, 73
Parapodia, 67-68
Parasites, 35
Parchment worm, 69
Pinworms, 52-53
Planaria, 11, 23-33
    anatomy, 23
    cave dweller, 25
    collecting, 11
    ears (auricles), 25
    eyes, 24-25
    feeding mechanism, 27
    land planaria, 25
    learning, 32
        memory, 33
        planaria research, 32
        RNA and learning, 33
        training, 32
    locomotion, 24
    mouth, 26
    pharynx, 26
    regeneration, 27-33
    sensitivity, 26
Plant traps, 53-54
Platyhelminthes, 4, 23, 35-41
Pollution and man, 77
Pollution and worms, 79
Polychaeta, 5
Polyclads, 19

Radial symmetry, 17
Reproduction of worms, 7
Ribbon worms, 35
Ringworm, 7

(83)

Roundworms, 5, 45-54
   anatomy, 45, 47
   classes, 5
   collecting, 14, 15, 46
   control of trichina, 49
   hookworm, 49
   RNA (ribonucleic acid), 33
   in soil, 14, 45-46
   species, 47
   Trichinella spiralis, 47-49
   Trichinosis, 47
   Vinegar eels, 46-47

Sandworm, 5, 67-69
   anatomy, 67
   life cycle, 68-69
   locomotion, 68
   size, 68
Scale worm, 69
Schistosomiasis, 37
"School for planaria," 32
Sea mouse, 69
Seat worms, 52-53
Segmented worms, 5, 55-66, 66-75
Setae, 55
Silkworm, 7
Sizes of worms, 20
Somites, 55
Spherical symmetry, 17
Spirorbis, 69
"Swimmer's Itch," 37
Symmetry, 17-19
   asymmetry, 17
   bilateral, 17-19
   definition, 17
   kinds of, 17
   radial, 17-18
   spherical, 17-18

Tapeworm, 35, 41-44
   anatomy, 41
   beef tapeworm, 42
   characteristics, 23-24
   life cycle, 41, 42
   in man, 42
   treatment for infestation, 43
Trematoda, 5, 35-41
Trichinella spiralis, 47-49
Trichinosis, 47
Tube-dwellers, 69, 71
Turbellaria, 5, 23, 25

Venus flytrap, 53
Vinegar "eels," 46-47

Worm "holidays," 72
"Worm manure," 65
Worm sleep, 67
Worms
   castings, 64, 71
   characteristics, 3, 9
   collecting, 9-10
   colors, 19
   containing chlorophyl, 19
   food for man, 20, 73
   habitats, 3
   luminescence, 21
   observation of, 10-11
   sizes, 20
   symmetry, 17-19
Worms made-to-order, 80

About the author

Katherine Nespojohn of Bridgeport, Connecticut, is fascinated by the world of physical science. She conveys her enthusiasm and interest to young readers in this absorbing trip through the "worm world," that dark, wet, and muddy underground which is home to many of these interesting creatures. Mrs. Nespojohn is a dedicated "worm-watcher," and encourages her young readers to become worm-watchers, too. This is the author's first book for Franklin Watts.